PROJECT MANAGEMENT FOR DESIGNERS AND FACILITIES MANAGERS

FOURTH EDITION

Howard G. Birnberg

Copyright © 2015 by Howard G. Birnberg

ISBN-13: 978-1-60427-120-1

Printed and bound in the U.S.A. Printed on acid-free paper.

10 9 8 7 6 5 4 3 2 1

Library of Congress Cataloging-in-Publication Data

Project management for designers and facilities managers /
By Howard G. Birnberg.—Fourth Edition.
   pages cm
Includes bibliographical references and index.
ISBN 978-1-60427-120-1 (hardcover : alk. paper) 1. Architectural firms—United States—Management. 2. Engineering firms—United States—Management. I. Title.
NA1996.B5 2015
720.68—dc23
                              2015013270

This publication contains information obtained from authentic and highly regarded sources. Reprinted material is used with permission, and sources are indicated. Reasonable effort has been made to publish reliable data and information, but the author and the publisher cannot assume responsibility for the validity of all materials or for the consequences of their use.

All rights reserved. Neither this publication nor any part thereof may be reproduced, stored in a retrieval system, or transmitted in any form or by any means, electronic, mechanical, photocopying, recording or otherwise, without the prior written permission of the publisher.

The copyright owner's consent does not extend to copying for general distribution for promotion, for creating new works, or for resale. Specific permission must be obtained from J. Ross Publishing for such purposes.

Direct all inquiries to J. Ross Publishing, Inc., 300 S. Pine Island Rd., Suite 305, Plantation, FL 33324.

Phone: (954) 727-9333
Fax: (561) 892-0700
Web: www.jrosspub.com

DEDICATION

To my wife, Diane Mix Birnberg and to my son, Michael Birnberg

CONTENTS

Preface xv
Acknowledgements xvi
List of Cases xvii
List of Figures xix
About the Author xxi
Web Added Value™ xxii

Chapter 1 Project Management Concepts 1
Introduction 1
   General Comments 1
   Design Firm Project Management 2
   Facilities Project Managers 3
   Project Management in Small Design Firms 3
   Defining the Project Team 4
   Problems with Weak or Ineffectual Project
      Management Systems 5
Project Delivery Systems 7
   System Types 8
      Pyramid Approach 8
      Departmental Organization 9
      Matrix Management 10
   Other Concepts 12
      Account Managers 12
      Studios 12
   Impact on Construction Costs 12
   Client Selection of Design Consultants 12
Strategic Project Management Plan 13
   Elements of a Strategic Project Management Plan 14
   Items to Consider under the Plan 15
Project Delivery Methods 16
   Traditional Straight-line Method 16
   Fast-track Method 18

v

Design/Build Method 19
Case Study: A Showpiece-turned-sour Triggers Change for
  Future Jobs 20

**Chapter 2   Planning and Management Concepts for
  Project Managers** 23
Introduction 23
   General Comments 23
   Planning Concepts 23
   Long-range Planning Process 24
   Goal Setting 25
Management Concepts 26
   Decision Making 26
   Roles of a Project Manager 29
Summary 30

Chapter 3   The Project Manager 31
Who Is a Project Manager? 31
   Survey Findings 31
   Characteristics of Project Managers 32
   Communication Skills 33
Summary 34
Case Study: Ohio Operating Power System 34
The Project Manager 36
   Areas Requiring Attention 36
Project Manager Responsibilities 38
   General Comments 38
   Specific Responsibilities 38
   Design Firm Project Managers and Marketing 40
Caring for and Training Your Project Managers 42
   Finding Project Managers 42
   Keeping Your Managers 44
   How Many Project Managers Do You Need? 44
Case Study: AB&C Telecommunications 46
Rewarding Project Managers 48
   Positive Rewards 49
   Training Project Managers 50
   Developing Your Training Program 52
   Staff/Management Training 52
   Determining if Training Is Needed 53
   Staff Development Officer 56

What Generates Effective Learning?........................ 57
Cross Training... 57
Training Practices and Methods 58
On-the-Job Training 59
Mentoring Programs....................................... 59
Program Evaluation.. 60
Managing the Training Program 60
Sources/Providers... 60
Mentoring Project Managers.................................... 61
Mentoring Programs....................................... 61
Types of Mentoring Programs.............................. 62

Chapter 4 Personnel Planning and Management............... 65
Personnel Planning 65
General Comments... 65
Leveling Workload 68
Contract Staffing 69
Contract Staff.. 69
Employers... 71
Summary 71
Staff Management.. 72
General Comments... 72
Techniques for Proper Delegation 72
Responsibility and Authority 74
Delegating Responsibility 74

Chapter 5 Soft Skills for Project Managers 77
Introduction 77
General Comments... 77
Time Management 77
Meeting Management....................................... 79
Other Ideas... 80
Telephone Time Management................................ 80
Telephone/Voice Mail Lesson.............................. 82
Preparing and Editing Written Materials...................... 82
Introduction 82
Public Speaking Techniques for Project Managers 85
Suggestions.. 85
Specific Suggestions to Improve Your
    Public Speaking Ability............................. 85
Summary 87

Successful Negotiating Skills for Project Managers 88
    Introduction 88

Chapter 6   Design Firm Operations 91
Introduction 91
Profit Planning for Design Firms 91
    Labor 92
    Non-Labor Costs 92
    Ratios/Multipliers 94
Scope Determination by Design Firms 95
    Dividing Contracts 97
Selecting External (to the Prime) Consultants 97
    Consultant Selection Process 97
    Owner/Facilities Managers Selection of Design Consultants 99
Budgeting Project Design Costs 99
    Project Cost Plan 102
    Direct Personnel Expense 105
Case Study: Federal Design Work 107
    Background 107
    MT&Y 109
    Argonne National Laboratory 111
    The Project 112
Other Issues 113
    Revealing Salary Information to Project Managers 113
    Value Pricing 113
    Reimbursable Markups 114

Chapter 7   Managing the Design Process 117
Case Study: Selecting the Right Team 117
Design Theory in Brief 119
    Changing Ideas 119
Evaluating and Selecting Designers 120
    What Clients and Facilities Managers Should Consider 120
    Living Together 121
The Design Process in Brief 123
    Parts of the Design Process 123
    Engineering Design 124
Summary 125

Chapter 8   Project Phases and Personnel Responsibilities 127
Project Phases 127

Introduction 127
Design Firm Staff and Responsibilities 130
    Position Descriptions 130
    Specialized Consultants 132
    Specialists and Generalists 133
Designer/Client Relationships................................... 135
    Client Retention... 135
    Owner/Client Expectations 136
    What Owners Should Ask Designers before
        Selecting Them.. 138
    Successful Projects Require Assertive and
        Knowledgeable Owners and Facilities Managers.......... 139
    Summary 144
Owner Program Management Options............................... 144
    Issues in Selecting Delivery Options....................... 145
    Owner Requirements 145
    Owner Capabilities... 145
    Owner Responsibilities..................................... 146
    Delivery Options 146
    Program Management: Extension of the Owner's Staff 147
    Program Management: Scope of Services 148
    Summary: Program Management Applicability................. 149

Chapter 9   Contract Management/Project Administration........ 151
Before Signing the Design Services Contract 151
    General Comments.. 151
Financial Issues 152
    Summary 154
Contract Types 154
    Types of Design Services Contracts 154
    Summary 156
Billing and Collection 156
    General Comments.. 156
Prompt Payment of Consultants................................... 158
    Issues 158
    How the Owner/Client Is Impacted.......................... 161
    What Can Owners/Clients Do? 161
Case Study: LOOP Architects 162
Scope Management.. 162
    Key Concepts 162

Design Contract Change Orders. 165
Communicating Design Change Orders. 166
Controlling Project Design Costs and Schedules 169
The Challenge 169
Why Projects Run Over Their Design Budget 172
General Comments. 172
Turning Problem Projects Around. 173
Project Administrative Activities. 174
Filing Project Paper Data 174
Incoming Correspondence 175
Outgoing Correspondence 175
Interoffice/Intraoffice Memos. 175
Project Meeting Notes 176
Work Authorizations. 176
Telephone Calls 176
Manufacturer's Assistance. 176
Confidentiality 177
Dealing with Consultants. 177
Checklists 177
Check Prints 178
Email/Text Messages. 178
Project Notebooks and Project Management Manuals 179
Project Notebooks 179
Project Management Manuals. 180
Development 180
Content. 180
Case Study: International Potato Corporation. 182
Background. 182
Facilities Management 183
Partnering 184
Concepts. 184

Chapter 10   Managing Project Quality and Risk Management . . . 187
Quality and Risk Management Concepts 187
Introduction 187
A Total Approach 188
Effective Communications 189
Documentation. 191
Summary 192
Quality Management 192
Introduction 192

Total Quality Management........................... 193
Developing a Quality Assurance Program...................... 194
    Steps in Developing a Quality Assurance Program............ 195
        Organization Plan..................................... 195
        Project Manager System 196
        Quality Assurance Development 196
        A Lawyer as a Team Member 196
    Summary 197
Peer Review 197
    Introduction 197
    Components of a Peer Review 198
    Peer Review: Help or Heartburn?........................ 198

**Chapter 11   Project Cost Control/Specifications/Value
Engineering** 201
Project Cost Control.. 201
    Introduction 201
    Controlling Internal Project Costs 201
    Communicating with Clients 202
    Information Systems.................................... 203
    Estimating and Controlling Construction Costs 203
Specifications 205
    Introduction 205
    Other Specification Issues 207
        Allowances 207
    Performance-Based Codes and Standards 209
Value Engineering... 209
    Introduction 209
    Implementation Results 210
    Involve the Customers 211
    Function Analysis 211
    Adhere to the Value Engineering Job Plan 214
    Conclusion 215

Chapter 12   Scheduling 217
Project Scheduling 217
    Introduction 217
    Scheduling Methods.................................... 217
        Full Wall Scheduling................................... 218
        Bar Charts.. 218
        Critical Path Method 218

Benefits and Limitations. 224
Issues in Planning and Scheduling for Project Managers. 226
    Introduction 226
    Defining Successful Project Completion. 226
    Purpose of Planning 227
    Planning Activities 228
    Why Schedule?. 228
    What Constitutes a Good Schedule? 228
    Products of a Schedule 228
    Responsibilities of Project Managers in the Preparation
        of Schedules 229

Chapter 13   Computer Applications. 231
Introduction 231
    A Brief Journey through Time 231
        The 1950s. 231
        The 1960s. 232
        The 1970s. 233
        The 1980s. 234
        The 1990s. 235
        The 2000s/2010s 236
What Do CADD Systems Do?. 237
    General Comments. 237
Web Collaboration Tools—Will They Fit on the
    A/E/C Industry Tool Belt? 240
    Introduction 240
    The Information Technology Promise. 240
    Flexible Project Delivery Systems. 242
    Site Administration and Information Control. 242
    Accelerated Decision-making Process. 243
    Web Collaboration Potential. 245
Building Information Modeling: Promise Unmet?. 245
    Historical Background. 245
Case Study: A Cautionary Digital Tale 249

Appendix 1   Project Closeout. 251
Introduction 251
    General Comments. 251
    Project Data Retention. 251
    Legal Concerns. 253
    Management and Marketing Use 254

    Project Managers and Marketers 254
Appendix 2   Project Environmental Considerations 255
Introduction 255
    General Comments. 255
The Green Building Revolution 256
    Electrical Generation 257
    Geothermal Energy. 257
    Water Harvesting Systems 258
    Costs/Benefits of Green Buildings 259
    LEED 259

Selected Bibliography 261
Books and Manuals. 261
Other Selected Resources. 264
Articles and Reports 264

Index. 267

PREFACE

Construction industry project management is an extremely broad subject ranging from detailed technical areas to intuitive people skills. Project managers (PMs) must be effective in dozens of skills if they are to successfully lead a project from conception to occupancy and beyond. Unfortunately, few individuals enter the design and facilities management professions to become project managers. Most professionals are interested in the creativity, challenge, and rewards of the design and construction process. Few engineering, architectural, or facilities management academic programs offer training in project management. As a result, most project managers learn their skills while on the job and achieve their management position more by accident than by plan. While the concept of project management is widespread in the design and construction industry, effective performance by individual project managers is often haphazard. Many organizations lack any form of a project management plan to guide the selection, training, and care of PMs and fail to fully develop necessary tools and systems.

In a single word, the project manager's job is to *communicate*. Without effective communication, the performance of every member of the project team will suffer. This often leads to a problem project and confrontation, not cooperation. *Project Management for Designers and Facilities Managers, Fourth Edition* is intended to provide current and aspiring PMs and design and facilities management organizations with the tools to achieve effective project management.

ACKNOWLEDGEMENTS

The author would like to thank the following individuals and engineering firm for their help in preparing this and prior editions of the book.

Michael Birnberg, Figure 5.1, *Basic American English Grammar*

Richard Pearce, Chapter Eight, *Owner Program Management Options*

Gene Montgomery, AIA, Chapter Nine, input on *Project Administrative Activities* topics and Chapter Twelve, *What Do CADD Systems Do?*

Geotechnology, St. Louis, MO, Figure 10.1, *Red Flag Words*

Jeff Orlove, AIA, Chapter Ten, *Developing a Quality Assurance Program*

John Schlossman, FAIA, Chapter Ten, *Peer Review: Help or Heartburn?*

Jeff Lew, Professor, retired, Purdue University, value added material, *ISO 9000*

Len Robin, Chapter Eleven, *Specifications* material

Howard Ellegant, AIA, Chapter Eleven, *Value Engineering*

Tom Eyerman, FAIA, Chapter Twelve, *Critical Path Method* material

Brad Sims, Ph.D., Chapter Twelve, *Scheduling* material

Robert Schneider, PE, *Web Collaboration Tools* material

LIST OF CASES

Chapter	Case Name	Page
1	A Showpiece-turned-sour Triggers Change for Future Jobs	20
3	Ohio Operating Power System	34
3	AB&C Telecommunications	46
6	Federal Design Work	107
7	Selecting the Right Team	117
9	LOOP Architects	162
9	International Potato Corporation	182
13	A Cautionary Digital Tale	249

LIST OF FIGURES

Figure No.	Name
Figure 1.1	Pyramid approach
Figure 1.2	Departmental organization
Figure 1.3	Strong project management matrix
Figure 1.4	Project delivery systems basic elements
Figure 1.5	Project delivery systems traditional straight-line
Figure 1.6	Project delivery systems fast-track
Figure 1.7	Project delivery systems design-build
Figure 3.1	Number of project managers leaving
Figure 3.2	Project manager training topics
Figure 4.1	Personnel planning
Figure 5.1	Basic American English grammar
Figure 6.1	Sample profit plan (labor only)
Figure 6.2	Sample profit plan (with non-labor costs)
Figure 6.3	Scope of services planning form
Figure 6.4	Information for a completed project file
Figure 6.5	Project cost plan
Figure 7.1	Some external constraints on designers
Figure 8.1	Building project phases
Figure 8.2	Project phases
Figure 8.3	Contractual relationships
Figure 9.1	Design services billing checklist
Figure 9.2	Work authorization form
Figure 9.3	Consultant work authorization form
Figure 9.4	Project notebook outline
Figure 9.5	Project management manuals
Figure 10.1	Red flag words
Figure 11.1	Selection of products
Figure 11.2	Specification types table of comparisons
Figure 11.3	Value engineering job plan
Figure 12.1	Gantt chart

Figure 12.2   Activity ground rules
Figure 12.3   Sample network
Figure 12.4   Example network #1
Figure 12.5   Example network #2
Figure 12.6   Summary of the planning and scheduling process
Figure A1.1   ABC Design report card form

ABOUT THE AUTHOR

Howard Birnberg is president of Birnberg & Associates, a management consulting and association management firm. He is presently serving as an instructor in project management at the University of California-Berkeley Extension and at Embry-Riddle (Worldwide) University. For six years, he served as an instructor on project management in the Office of Executive Education at the Harvard University Graduate School of Design and also as an adjunct assistant professor at Michigan State University, College of Human Ecology. Mr. Birnberg formerly served as a lecturer on project management for the University of Wisconsin Department of Engineering Professional Development for nearly twenty years. He has also lectured at the University of Illinois-Chicago, Northwestern University, University of Texas-El Paso, University of Kansas and Andrews University.

He is the author of *Project Management for Designers and Facilities Managers, 3rd ed.* (J. Ross Publishing, 2008), *Project Management for Building Designers and Owners* (CRC Press, 1998), *Project Management for Small Design Firms* (McGraw-Hill, 1992) and served as general editor of *New Directions in Architectural and Engineering Practice* (McGraw-Hill, 1992). His research and articles regularly appear in professional journals, and he is currently serving as a columnist for *Civil & Structural Engineer* magazine. Mr. Birnberg holds a Bachelor of Science in Architecture degree from Ohio State University and a Masters in Business Administration from Washington University in St. Louis.

This book has free material available for download from the
Web Added Value™ resource center at *www.jrosspub.com*

At J. Ross Publishing we are committed to providing today's professional with practical, hands-on tools that enhance the learning experience and give readers an opportunity to apply what they have learned. That is why we offer free ancillary materials available for download on this book and all participating Web Added Value™ publications. These online resources may include interactive versions of material that appears in the book or supplemental templates, worksheets, models, plans, case studies, proposals, spreadsheets and assessment tools, among other things. Whenever you see the WAV™ symbol in any of our publications, it means bonus materials accompany the book and are available from the Web Added Value Download Resource Center at www.jrosspub.com.

Downloads for *Project Management for Designers and Facilities Managers, Fourth Edition*, include an overview of ISO 9000 quality standards and commissioning.

1

PROJECT MANAGEMENT CONCEPTS

INTRODUCTION

General Comments

Few activities are as complex as building design and construction. Compounding this complexity is the fragmented nature of the team charged with designing and constructing buildings and other facilities. In many owner/client organizations, the decision makers are often unaware of the complexities and costs of the process, and may make decisions without the input of those assigned to bring these to fruition. The *Great Recession* significantly impacted the construction industry. It took years for the industry to recover and it emerged much leaner than before. Workloads that began falling in 2008 continued for years, resulting in enormous layoffs of staff. Contractors, in particular, found many older workers opting for retirement and younger ones seeking work elsewhere. As a result, a labor shortage emerged, particularly in the project manager (PM) ranks. Because of the changed economy, engineers, architects, and facilities managers have been required to learn new skills and reposition themselves. Traditional delivery approaches have been altered, and new ones emerged. The pre-recession narrowing of the need for some services, such as traditional architectural services, has continued unabated. Services previously handled in-house in larger owner/client organizations are now being outsourced, greatly reducing the in-house facilities design and construction staff members who are available to plan, coordinate, and manage owner/client projects. Some of these outsourced specialists are traditional engineering, architectural, and facilities firms—others are found in niches such as program and project management firms.

Many design firms now face challenges in finding and keeping capable PM staff, in developing better project scope management systems, in complying with fierce fee and time pressures, along with a myriad of other issues. The *Great Recession* resulted in many experienced people leaving the industry, some retiring early, or young people choosing other fields. As a result, there is a gap in the ranks of capable PMs. Identifying and training PMs is a pressing need in the construction industry.

Outside of the construction industry, the term *project management* often broadly means *scheduling*, and *project management software* is simply *scheduling software*. Inside the industry, that is only one tiny fragment of the package. However, the entire concept of using PMs is borrowed from outside our industry, and their fundamental role in managing and controlling the project scope, schedule, and budget remains the same both within and outside of the construction industry.

To achieve their goals, design and owner/client PMs must clearly understand their primary role. They are to serve as the communications conduit in a highly fragmented, specialized, and complex undertaking. Their tools must allow them to communicate quickly and effectively.

Design Firm Project Management

A fundamental element of design firms is their project management system. This system enables firms to complete projects successfully and hence solve their client's problems. The project is the profit center of a design organization. The PM is in the best position to control the final outcome of a project and can have a great impact on project and firm profitability. However, not all PMs function in the same manner. Many design firms have what is known as a weak project management system. Typically, in this type of system the principal or partner is responsible for actually producing the work. In effect, the PM is really a technician.

I have often heard design firm practitioners say that profitability and growth are factors of the ability to get the work (success of the marketing effort). I believe that this is only partially true. A much more fundamental factor is the firm's project management system. With a strong project management system, design firm principals can and should spend more effort toward getting the work and managing the firm.

Many design firms pride themselves on personal service to clients. However, with a weak or ineffective project management system, the principals often deal with the client, and the PM is in charge of producing the work. No single individual has complete control of the project, hence efficiency and profitability suffer. In essence, when more than one individual is responsible for the project, in reality no one is. As a result, the desired personal service is often a myth because those who are actually responsible for performing the work rarely communicate directly with the client.

Facilities Project Managers

There is no single model for project management in facilities organizations. This is a direct result of the varied nature of these clients. In general terms, clients include the commonly accepted groupings of public, private, institutional, governmental (essentially the same as *public*), corporate, institutional, commercial, residential (high-rise, low-rise, single family, apartments, etc.), retail, and many other categories. As found in design firms, clients can be large or small, domestic or international, specialized or generalized, or fall into many other groupings. Some facilities staff might consist of a handful of people; others might number in the thousands—some are centralized; others dispersed to many locations. Their roles vary and can be ever-changing. I have seen some reorganize regularly; others have decided their model works fine and haven't changed it for years. Many of these static facilities groups are poorly equipped to meet the needs of their organizations when internal or external pressures require a different response. Two case studies in this book contain elements of this situation (*AB&C Telecommunications* in Chapter Three and the *International Potato Corporation* in Chapter Nine).

Despite the lack of a clear model for facilities project management, there are many common skills, tools, systems, and approaches needed by all. This book presents many of these necessary for both design and facilities PMs. Essentially, good communication is at the core of all effective project management in the construction industry.

Project Management in Small Design Firms

Effective project management is important to all organizations. Small design firms are especially handicapped by a lack of project control and reporting systems, a lack of staff time solely devoted to managing projects, insufficient principal time to run both the firm and projects, and an inability to market with enough regularity to ensure a steady workload.

In addition, many smaller design practices find it difficult to apply published or seminar material on project management. The situations described often assume that the firm has the necessary staff and resources to implement prescribed systems and activities. In many small design firms, principals run the firm and its projects. They make all decisions (major and minor), meet with clients, keep project and financial records, negotiate contracts, and are involved in dozens of other tasks on a daily basis. Many of these individuals find it difficult to understand how project management techniques that are presented for larger design organizations are relevant to their situation.

There are however, many excellent procedures appropriate for large and small organizations that enable firms to escape the treadmill of constant crisis management. Systems and methods can be implemented that will make design practice more efficient and profitable. For example, firms with projects of shorter durations may want to selectively incorporate project management techniques based upon the specific situations presented. It may not, for example, be sensible to establish an elaborate project status reporting system for projects whose duration may be measured in days, rather than months or weeks. In this case, the system may not be current enough to allow the principals or PM to take corrective action in the event of problems. The reporting system should, however, provide enough information to record historical data on project profitability, change orders required, basic contract terms, and similar information.

Defining the Project Team

Many organizations and PMs narrowly define a project team. Most consider the project team to only include those immediate firm members working on a particular project. The project team actually should include members of all organizations involved with a project—including clients, consultants, suppliers, contractors, subcontractors, engineers, architects, and so on. The total number of team members may include dozens or even hundreds of individuals from many different organizations. Each individual team should be represented by a PM who is the permanent representative to the overall project team.

Problems with Weak or Ineffectual Project Management Systems

Many firms believe they have effective project management systems, but these are often flawed. There are many signs which point out a weak or ineffectual system—several are discussed below. While no system is without its flaws, the goal of any firm manager should be to minimize their impact. Unfortunately, the pressure of meeting project requirements often results in necessary changes being pushed aside by the notion that it is more expedient to continue operating as you always have. Eventually, these inefficiencies and weaknesses can result in reduced profits, poor service, quality issues, and many other undesirable consequences. An important concept to consider is the need for continuous improvement in your operations. Even effective project management systems require regular fine tuning to incorporate new ideas and concepts, make the best use of new technologies, adapt to changing needs and economic conditions, develop new services, and provide the staff with the skills they require.

1. **Overburdened senior managers:** Weak or ineffective project management systems typically rely on a handful of senior managers to run the organization and manage projects. Under the umbrella of project management, they often market the firm's services (design organizations), develop a scope of services and project design budgets, prepare a project program, select outside consultants and providers, organize schedules, review work, and handle dozens of other activities. They also need to monitor the project scope to determine when and if activities lie beyond the range of services covered under the design contract. In the event that these activities are outside of the agreed scope, they need to alert the client and negotiate an increase in the design fee, if appropriate. Overburdened managers often lack the time or focus to handle these *scope creep* situations, resulting in penalizing the project's budget and profitability.
2. **Poor internal communications:** The impact of overburdened senior managers has a snowball effect on the most important aspect of effective project management *communications*. Managers lacking the time, focus, or systems to communicate project information to team members will find they need to redouble their own efforts to mitigate the consequences of this failure. Complicating the situation is the lack of a firm-wide approach to sharing infor-

mation, making decisions in a timely manner, and the need for staff to adapt to the unique style and approach of the various senior managers. Often distracted by more global firm management and/or marketing issues, senior managers may fail to transmit vital data to the project team members. Crisis management often prevails in these organizations.

3. **Poor external communications:** Many of the same problems highlighted in Point 2 impact those outside of the design or facilities organization. Clients, consultants, vendors, suppliers, contractors, subcontractors, and many others find themselves lacking necessary or timely information. This can result in project delays, higher costs, errors, and many other problems. While a highly effective project management system isn't always a panacea, it provides a mechanism to deal with these issues that is lacking in weak or ineffectual systems.

4. **Lack of decision making by those holding the nominal title of project manager:** Some organizations have a loose definition of the term *project manager* and assign the title to many people. Unfortunately, some of these individuals have many of the responsibilities of a PM without the necessary authority to make decisions. Authority often resides with senior managers who may not hesitate to step in when they have different priorities, concepts, and goals than the PM. They may contradict decisions made by the PM, meet with clients without the PM present (design firms), give instructions to staff, meet with consultants, or do a wide variety of activities undercutting the nominal PM. As a result, the PM often ceases to be effective, stops making decisions, and reverts to the role of *doer*, rather than manager.

5. **Lack of decision making by clients:** To be successful, projects require an active and assertive client who is willing to provide input and make decisions. Clients or their facilities representatives must make themselves available to the project team and quickly respond to questions. This is a difficult process in the best of circumstances and nearly impossible without an effective project management system where a PM anticipates the need for client input. Design PMs must regularly communicate with clients or their facilities representatives and push for decisions. Clients must be provided with complete and accurate information on a timely basis. Weak or ineffectual systems lack this important mechanism to ensure client decision making.

6. **Exceeding a manager's *span of control*:** There have been many studies identifying the limits to a manager's span of control. This term refers to the limit on the number of people an individual can effectively manage by answering questions, reviewing work, providing direction, giving instructions, and dealing with problems, conflicts, etc. Beyond this point, a manager becomes overstretched and increasingly ineffectual. A firm with a weak project management system tends to rely on a very limited number of people to manage the firm and projects. On large project teams, a PM will need to have a cadre of assistant PMs to handle various aspects of a job to prevent exceeding their own span of control. In firms where senior managers try to manage the business and projects, the span-of-control limit typically sets how large the staff can become before the operation begins to break down. Studies note this limit to generally be around ten subordinates per manager.

There are many other problems that are apparent in an organization with a weak or ineffectual project management system. Symptomatic of these is the neglect of general planning, financial management, and long-term marketing and public relations. These important subjects often take a back seat to the more immediate crisis management needs of current projects. In situations where a nominal PM has responsibility much greater than their authority, morale can suffer, conflicts may arise, and the PM may seek other employment. There are many circumstances where senior managers complain that their PMs won't take responsibility on jobs. While this may be true in some cases, often this perceived poor performance may be the result of a senior manager preventing the PM from properly doing their job. It could be that this PM may not be the appropriate person for the role, may lack training, may lack necessary tools, or the system itself may be ineffectual.

PROJECT DELIVERY SYSTEMS

There are as many methods of organizing for project delivery as there are firms. Most well-managed organizations are forever tinkering with their systems in an effort to improve service and their own profitability and performance. Some firms even use different systems within their operation based upon the project size, location, client need, or other factors. However, these systems generally fall into one of three formats.

System Types

Pyramid Approach

This method revolves around a key individual who makes all major (and often minor) decisions (see Figure 1.1). It is commonly used in smaller organizations where the firm's owner has daily involvement in all project and firm management decisions. One or more key technical and administrative support people aid in executing required tasks. Lower-level staff members generally perform only assigned tasks. If the key individual is not available, activity often grinds to a halt. In a busy firm, daily crisis management may rule, and important planning and operational issues may be neglected.

In a design firm, when an organization has two or more partners, multiple pyramids may exist. Each partner may have his or her own client group and key technical assistant. Formal coordination between partners rarely exists and important planning issues are often neglected. Where one principal is more successful in obtaining work, his or her share of the firm's workload may become overwhelming. This often creates friction and conflict between partners. Some firms deal with these problems by organizing an informal division of labor whereby assigned project and administrative roles are given to each principal. Unfortunately, most principals still prefer project involvement and may continue to neglect their assigned administrative or marketing functions.

Client organizations are not immune to the issues raised in organizing for project delivery with pyramidal structures. For example, a developer of commercial or residential projects may be deeply involved in all or most decision making. A manufacturing company with only a handful of locations may assign facilities management to an executive who may also have other responsibilities. Managing projects for the company may be a part-time activity for which he or she may not be trained and may

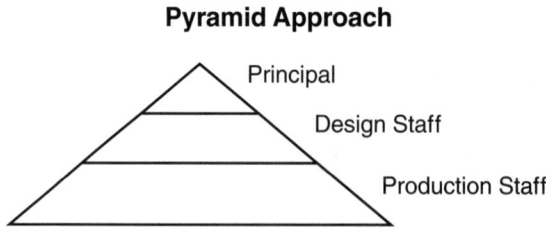

Figure 1.1 Pyramid approach

neglect facilities management. Smaller municipalities may have a mayor who has staked his or her reputation (and re-election possibilities) on the success of a project and may become deeply involved without the necessary skills or structure to be effective.

Departmental Organization

As organizations grow, pressure increases to formalize project and non-project management assignments. The division of labor concept often becomes the basis for a formal departmental structure. For example, a single-discipline engineering firm often structures itself around marketing, design, production, and field departments. Multidisciplinary firms may establish each discipline as a separate department (see Figure 1.2).

Single-discipline firms frequently appoint department heads drawn from associate or principal ranks. These individuals normally retain full authority for all project and non-project decision making within their departments. In theory, responsibility for project management is delegated to the next lower staff level, and these individuals are called *project managers*, *project engineers*, or *architects*. Typically, their role is focused on technical issues, and they have little actual authority.

The concept behind the departmental approach is that as one department completes its work on time, within the budget, and according to the contractual scope, project authority and responsibility is passed to the next department head. In reality, department heads determine project priorities based upon their own workload, level of interest in the project, relationship with the client, and other factors. Often, each department and its head have different methods of project management and capabilities. As a result, the client must adjust his or her operation to three or more different project styles and systems. In addition, if the firm's principals assume departmental responsibilities, firm management and marketing may be neglected. Under the single-discipline departmental organization, the situation may be little improved over the pyramid approach, as

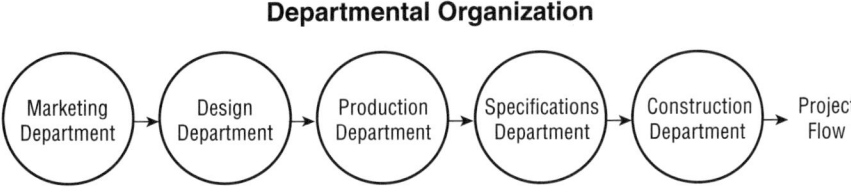

Figure 1.2 Departmental organization

each department may simply become its own pyramid. With no single individual in charge of the entire project, personal service to clients is often a myth and time schedules, budgets, profits, and the long-term well-being of the firm may all be jeopardized.

In an attempt to deal with the deficiencies of the single-discipline departmental system, some design firms have introduced the PM concept. Unfortunately, this system is often not designed or operated properly. All too commonly, firms institute a weak project management system where there is a significant imbalance between responsibility and authority. To be successful, levels of responsibility must be generally equal to levels of authority. In many firms, real decision-making authority remains with department heads while much of the project responsibility is delegated to PMs. As a result, PMs lack the authority, training, or experience to make decisions stick and simply cease being decision makers. This discredits the project manager/management systems in the eyes of department heads, senior managers, and clients. Where department heads are generally owners and PMs are associates or employees, the situation may become intolerable. In these firms, staff turnover may be high, profitability and productivity low, and client satisfaction questionable.

Multi-discipline firms can successfully use a departmental approach when many projects are within a single discipline. Frequently, these firms establish a matrix management system within each department. Where studios are used, a similar system may be developed. Multidisciplinary design firms with projects involving several disciplines require a project management system to handle work that crosses department lines. Often, a PM is selected based upon the predominant discipline required. Without an effective PM, multidisciplinary firms may find themselves with the same problems as single-discipline, departmentally organized firms.

Matrix Management

For many design firms and owner/client organizations, the matrix management (or strong PM) approaches functions best (see Figure 1.3). Under this system the PM is in full charge of the project from beginning (marketing) to end (continued contact). The PM has an equal balance of authority and responsibility and serves as the primary contact point for other members of the project team. His or her major responsibilities include meeting the program, schedule, and budget while maintaining profitability for the firm (if a design organization). Department heads or

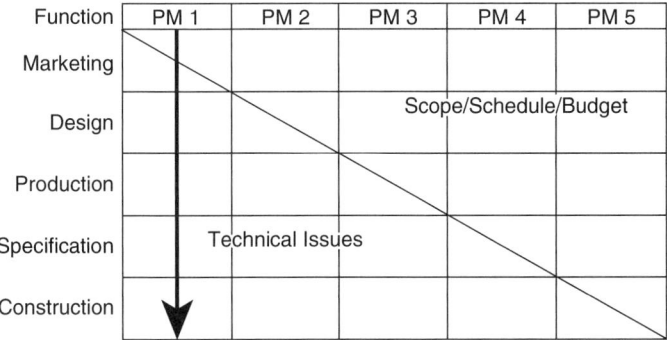

Figure 1.3 Strong project management matrix

chiefs of the various functional areas (e.g., marketing, design, production, etc.) retain responsibility and authority for technical decisions, staff assignments, training in technical areas, implementing quality reviews, and so on. Where technical decisions significantly impact the project scope or budget, the design firm and the client PM must review and approve decisions.

The matrix project management system separates the ownership role from the need to produce a successful and profitable project for both the client and the design firm. Unfortunately, at times of conflict, there is a tendency to confuse ownership with a project's real needs. To be successful, a firm using the matrix management system must be able to communicate to its staff and management how the system works. On occasion, confusion can arise on the part of staff members regarding whom to contact with questions, comments, or problems (the PM or a department/functional head). In addition, matrix management requires a complete, timely, and accurate job cost reporting system. Other management tools such as a project management manual are also extremely helpful.

Small design firms or large firms with small projects may establish an abbreviated matrix system. In this case, an individual may wear several hats and must be careful to focus on his or her specific responsibilities on a project. Some firms use the full matrix system only on their larger projects and use other approaches on their remaining jobs.

The matrix system requires a strong chief executive to arbitrate disputes between PMs, department/functional heads, and staff members. Although the goal of the system is to avoid bottlenecks and crisis management, disputes will occur and these require careful resolution. When functioning properly, matrix systems encourage decision making at the lowest possible level of the organization and both responsibility and authority must be delegated to make this possible.

Other Concepts

Some organizations use variations of the project delivery systems noted in this section. Two in particular require some attention.

Account Managers

Under this approach, instead of managing projects, a design firm PM manages a client account. This can be a successful approach, but often leads to imbalances in PM workload. A particular account manager may become overwhelmed with a very active client, while another may be under-utilized by slow client activity or temporary delays in funding, project approval, and so on.

Studios

Found in design firms, studios typically focus on a particular type of work such as healthcare, industrial/commercial, and interior design. Each studio is led by a senior manager or principal and is run like a semi-autonomous firm. Sharing of staff between studios is not well coordinated and can result in staffing imbalances and higher overhead costs. The benefit of studios is the ability to focus staff and expertise on a specific type of work. Normally a project management system should be employed within the studio.

Impact on Construction Costs

A poorly developed project management system can be costly to both a design firm and its clients. A design firm's ability to stay within a design fee or to manage design change orders (and often obtain additional fees) is dependent on their project management system. Construction costs can skyrocket when inadequate project management causes poorly prepared construction documents. Contractor bids may be high to cover unknowns or questions on design drawings, or low to obtain the job (then made up through construction change orders).

Client Selection of Design Consultants

Among the factors that clients and their facilities staff should consider in the selection of design consultants is the designer's internal project delivery system. It is usually best to select a consultant whose system compliments their own. For example, if a client organization has a strong project management system, selecting a departmentally structured design consultant may result in a frustrating and unproductive relationship.

STRATEGIC PROJECT MANAGEMENT PLAN

Project management is not a static activity. Developing and improving your project management program requires a plan, regularly updated and directed. This strategic plan guides your decision making regarding hiring, training, software acquisition, project management tools development, organizational changes, compensation, and dozens of other topics.

As is true with all strategic planning efforts, the strategic project management plan requires a step-by-step approach to develop and maintain successfully. These steps typically include an assessment process, benchmarking of best industry practices, determination of goals and objectives, writing of a plan, implementation, and updating/revising. Unfortunately, in most design and construction organizations, the project management program evolves without any planning. Often, it is growth and high workloads that force the change to some form of project management. Typically, project management systems in these organizations lack coherence, the necessary tools, and capable managers.

The typical evolution of organizations in the construction industry includes several steps. Embryonic or early stage firms are typically pyramidal in nature with the owner/founder at the top of the pyramid. This individual is both firm manager and PM. For many, their normal day is one of crisis management with little opportunity for planning, reflection, or change. Some escape this treadmill and develop growing organizations. If there are several individuals at a senior level (i.e., multiple owners and principals), then often a departmental organization evolves where each senior individual has his or her own segment of the firm to run. An example would be a single-discipline engineering firm where one partner is in charge of the design department, a second is in charge of the technical department, and a third handles marketing. While the departmental organization can work in multi-discipline firms, it typically fails to build a coherent project management system in single-discipline firms.

Some firms overlay a weak project management system on top of a departmental structure. This creates a situation where middle managers have most of the project responsibility while senior managers retain authority. This is often the case in governmental bodies and corporate organizations where the internal project management/facilities management group has little involvement in determining needs, budgets, schedules, and so on. Typically, these individuals have responsibility for meeting the needs of their internal client, but have little or no control over resources.

The next step in the evolution of design and construction organizations is the development of the matrix or strong project management system. When properly functioning, PMs in this system have relatively equal levels of authority and responsibility. They also have all of the necessary tools to do the job and are provided with opportunities for training and advancement. Mentoring programs exist to identify and educate future PMs and organization leaders. Reward programs are developed to provide incentives for excellent performance.

Many organizations claim they have a strong project management system, but an objective examination of their system would find many failures and problems. It is astonishing that many in the construction industry who are professional planners by trade do so little for their own organizations. The typical project management system develops without a plan and lacks any guidelines for the future.

Elements of a Strategic Project Management Plan

What are the elements of a strategic project management plan? The following highlights some of the most important items to include.

1. **What is the current status of your project management system?** How is it performing? Are you providing the best service to your internal and/or external clients? Do you have excessive turnover in the ranks of PMs? Do you have all necessary project management tools? Do your PMs get information in a timely fashion? Dozens of similar questions must be answered objectively.
2. **What are the industry benchmarks for best practices?** Do you know how project management is handled elsewhere in the industry? In each area of your project management operation, what are the best practices used by others that could benefit you and your internal and/or external clients?
3. **What are the goals and objectives for your system?** For example, is your system a training ground for future owners of the firm? Is your goal for project management to maximize profits, provide

top-notch service, or both? Is the use of cutting-edge technology vital to your organization? Do you want a culture of continuous improvement built into the system (which can be chaotic)? Do you want to structure career paths for PMs (and possibly all staff)?

4. **Prepare a written plan.** Document all items mentioned in Points 1–3. Prepare an action plan. What should you do first? Second? Provide a budget and resources and include them in the plan. Set a schedule and follow it. Share the plan with all staff, not just PMs. Most of all, keep it simple and achievable while making it challenging enough to require real effort on the part of your organization.
5. **Implement your plan.** This is the hard part. Planning is tough enough. Implementation is challenging. It requires action. You need to spend time, money, and effort. Most organizations achieving success at implementation establish a planning committee to oversee the process. This group meets regularly and makes ongoing decisions under the guidance of the strategic project management plan.
6. **Update and revise your plan.** Changes to the plan are made when a need is determined. It is not a stagnant document and may require occasional revision. The planning committee should periodically examine and update your plan with input from those affected by it.

Items to Consider under the Plan

There are potentially hundreds of issues to consider under the strategic project management plan. The following lists some of the most common items to examine.

1. PM compensation
2. PM incentives
3. Reporting tools
4. Computer systems: CADD, BIM, job cost reporting systems, technical software, scheduling tools, budgeting tools, database tools, enterprise software, etc.
5. Support staff: accounting, administrative, etc.
6. Office facilities
7. Communication tools: cell phones, laptops, tablets, etc.

8. General project management organization
9. Non-computer tools: project notebooks, change order management (design and/or construction, budgeting systems)
10. Roles and responsibilities of PMs
11. Training program
12. Mentoring program
13. Defining your ideal PM
14. Communication channels (existing and desired)
15. Organizational growth prospects
16. Sources of project management talent
17. Assessment of current PMs
18. Profits? Service? Quality? All?
19. Project management manual
20. Other

PROJECT DELIVERY METHODS

In the construction process, there are three primary methods of project delivery: (1) traditional straight-line, (2) fast track, and (3) design/build. No single method is superior, in itself, to the others. The primary factors which influence the use of one method over another include time frames, cost constraints, and quality. Other project delivery methods could be identified; however, many are only variations of those listed.

Traditional Straight-line Method

The process begins with the designer determining the client's (owner's) physical and functional needs. They then follow through with the creation of the building design, obtain the necessary project approvals, produce all required drawings, obtain either competitive bids, or negotiate prices with contractors before construction begins (see Figures 1.4 and 1.5). This flow allows complete development of the design program, fully developed design solutions, and complete pricing of building costs prior to construction.

This is the method most familiar to designers (particularly architects) because it blends smoothly into their operation. With this method, personnel needs can be planned, and it allows for the use of traditional management methods. Generally, in this method of project delivery, the designer has greater input into the project program than with other methods.

Project Delivery Systems
Basic Elements

| Design and Documents | Bidding and Negotiation | Construction |

Figure 1.4 Project delivery systems basic elements

Project Delivery Systems
Traditional Straight-line

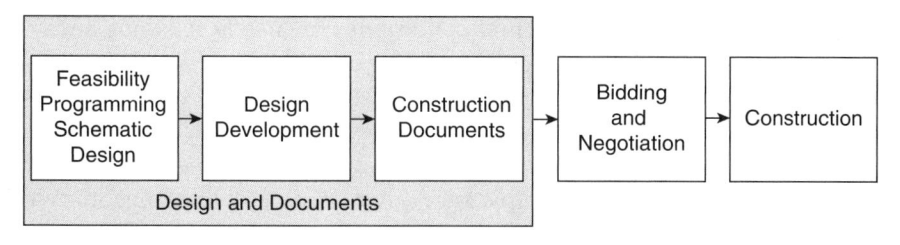

Design and Documents: [Feasibility Programming Schematic Design] → [Design Development] → [Construction Documents] → [Bidding and Negotiation] → [Construction]

Figure 1.5 Project delivery systems traditional straight-line

For private sector owners, pricing is often based upon a lump-sum general contract, under which the contractor constructs the client's building for a fixed price based upon the designer's drawings and specifications. The contractor will (or should have) performed a complete analysis of costs for the project, including subcontracts, special equipment requirements, and materials. If the contractor properly completes his or her estimate, a residual (profit) should remain upon completion of the building. In the event that he or she exceeds this fixed amount, there usually is no recourse other than renegotiation of the contract or lose money on the job.

It is not uncommon for the client and/or designer to retain outside cost consultants to monitor project costs during design, or to monitor the accuracy and fairness of the bids or the contractor's negotiated price. The deliberate nature of the traditional straight-line method allows opportunity for the client to increase the budget if necessary, cause redesign of the project, or reduce the project scope. Particularly with public sector owners, this process is vital since the budget is often set by appropriation and it is not always easy to obtain additional funding.

If the contractor is locked into a lump-sum contract and the project costs exceed this amount, the contractor is subject to a loss. Many prefer the use of cost-plus contracts. In this method, the contractor is reimbursed for all project costs and is paid a fee in addition. A pure cost-plus contract reduces the contractor's incentive to keep project costs down; hence, the use of *upset maximums* (at a percent over the original estimates) is widespread. Beyond the maximum, the financial burden of cost overruns shifts back to the contractor. Frequently, however, there is an agreement between the owner and contractor to split all savings under the upset as an incentive for the contractor to control and/or reduce costs.

Fast-track Method

With quickly changing economic conditions and needs, it is not always possible to wait for completion of a building under the traditional straight-line method of project delivery. In response, the fast-track method was developed to allow for the overlap of the design and construction phases. Upon completion of the project program and of the schematic design, the project is divided into bid packages that are awarded (or negotiated) in order by construction logic so that the earlier packages can be in construction while later packages are still in design (see Figure 1.6).

The impact of the fast-track method on designers and their PMs can be great. It increases the demand for individuals with specific skills considering that the needs of construction imperatives and costs require early decision making in key areas (such as mechanical systems that often

Project Delivery Systems
Fast-track

Design and Documents		
Bid Package A →	Bid Price for A →	Construction
Bid Package B →	Bid Price for B →	Construction
Bid Package C →	Bid Price for C →	Construction
Bid Package D →	Bid Price for D →	Construction

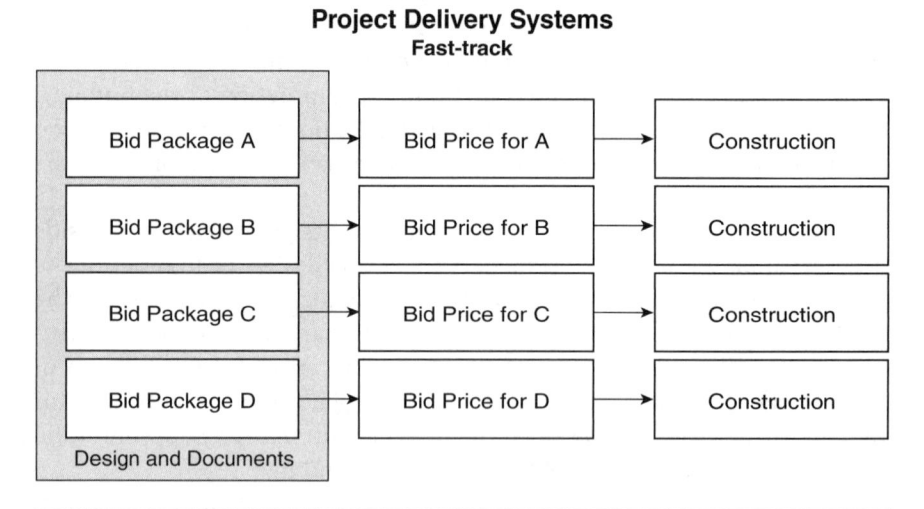

Figure 1.6 Project delivery systems fast-track

require long lead times for their manufacture). The effect of these decisions on later bid packages must be anticipated. Input to the design firm by the building product manufacturer or supplier, the general contractor, and the construction manager (CM) is often valuable to the success of this method. This input may be in evaluating product performance, costs (initial and life-cycle), and delivery times. Often, contracts are issued on an installed basis with a performance specification again requiring a manufacturer's input.

The use of multiple bid packages (contracts) and the general complexity of many projects have required the hiring of a CM to ensure the performance of the contractor(s) and the accuracy of both cost estimates and schedules. In this sense, the CM performs the function carried out by the general contractor on less complex projects. As mentioned earlier, the CM often functions as the general contractor; however, any qualified party (e.g., architect or engineer) could be the CM.

In general, there are two types of CMs. *At-risk CMs* typically provide a guaranteed price to the owner based upon a documented scope of services and building program. Their profit is derived from completing the project below the guaranteed price. As a result, they have a great incentive to control the fees and costs of design consultants, contractors, and others involved in the project. *Not-at-risk CMs* are typically hired by an owner to manage the construction phase activities on a project. Traditionally, architects handled this activity; however, many have withdrawn from providing the service due to liability, lack of fees, and other concerns. As a result, CMs have embraced the opportunity.

Design/Build Method

The use of the term design/build has come to be an umbrella—covering many variations of the same process. Basically, this method of project delivery involves quoting the owner a price at an early stage for both design and construction of the project (see Figure 1.7). Given the limited nature of the building program at this point, there is a great burden placed on the design/build firm to be able to accurately identify general building components and costs. The essential identifying element of the design/build method of project delivery is the single point responsibility of the design/build principal. This principal may be the architect, engineer, general contractor, specific design/build firm, or other consultant.

Often, there are a great number of variables in the selection of materials, equipment, and systems. This offers many excellent opportunities for the design/build team, as any cost savings achieved during construction

Project Delivery Systems
Design-build

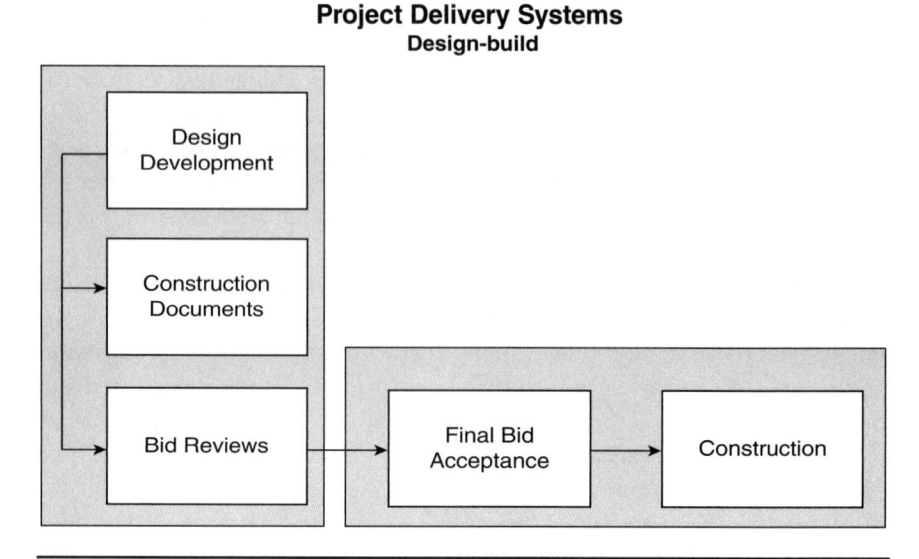

```
┌─────────────────────────┐
│  ┌──────────────┐       │
│  │    Design    │       │
│  │ Development  │       │
│  └──────────────┘       │
│                         │
│  ┌──────────────┐       │
│→ │ Construction │       │
│  │  Documents   │       │
│  └──────────────┘       │
│                ┌────────┼──────────────────────────────┐
│  ┌──────────┐  │ ┌──────────┐      ┌──────────────┐    │
│→ │   Bid    │─→│ │ Final Bid│  →   │ Construction │    │
│  │ Reviews  │  │ │Acceptance│      │              │    │
│  └──────────┘  │ └──────────┘      └──────────────┘    │
└────────────────┼──────────────────────────────────────┘
```

Figure 1.7 Project delivery systems design-build

are beneficial to the design/build contractor. Some design/build teams are true single entity organizations encompassing both design and construction groups. Many however, are cobbled together by an architect or contractor on a project-specific basis simply in response to a market opportunity. In general, they should still be considered two separate organizations. An owner should view this type of design/build team with care as coordination, cost control, and communication may be weak.

CASE STUDY

From: ***ENGINEERING NEWS-RECORD,***
May 11, 1998, Stephen H. Daniels, author

"A Showpiece-turned-sour Triggers Change for Future Jobs"

(Reprinted courtesy of Engineering News-Record, Copyright Dodge Data & Analytics 1998, All Rights Reserved)

On April 16, nearly a year late and $20-million over budget, University of Washington officials finally dedicated a still somewhat incomplete

$98-million engineering building—a 260,000-sq.-ft. showpiece turned sour.

The original plan was for a five-year, two-phase project. But phase two, renovation and reconstruction of an existing engineering and computer science building was scrapped after four years. Only a termination agreement between the university and its subcontractors saved some from financial ruin, says Earl Dutton, owner of Dutton Electronic Co., Inc., Seattle, one of the job's subcontractors.

The University of Washington's building binge embraces more than $600 million in construction since 1990. "The university has had a very good track record compared to many public owners," says Connie Miller, assistant vice president for capital projects. "Most of our projects have come in on time and on budget." However, Miller says this one "was awful".

The reasons are manifold and classic. Despite formal partnering, just about everyone involved says the project suffered from a "lack of coordination" between engineers and architects. Walls were built and torn out. Mechanical and electrical subcontractors, working from flag notes on incomplete blueprints, butted up against one another. Lack of detail in working drawings meant that mechanical subs were on their own to figure out configurations. For example, Dutton ran communications wiring outside because it would not fit inside.

The University blames its architect, Boston-based Kallmann McKinnell & Wood, for many of the problems. Kallmann principal, Henry Wood, in turn blames the university. He says that in a rush to get bids out in late 1993, the school went to bid with incomplete design drawings, then ordered more last-minute changes than the design team could accommodate.

Miller admits that as the Seattle construction market began to heat up, there was considerable pressure to get the project "to the streets." She says: "We talked with the design team. We thought we were ready to bid. The bottom line, simply, is we were not."

Wood also fingers the general contractor, Ellis-Don Construction, London, Ontario, saying it low-balled the conventionally bid job, then sought change orders to make a profit.

Bruce Blair, Ellis-Don's vice president and project executive, says that's not true. He explains that after seeking owner authorizations for architect-initiated or other changes, "we asked for money to do the assigned work." Blair adds that the firm had to be diligent toward that end.

Miller says it was clear that trouble was brewing by mid-1994, "six months into the work." By autumn, subcontractors and the general contractors began falling behind. Field problems were "severe," she says. "Every major subcontractor struggled."

Taken one by one, the snags, like recognition late in the game that elevator shafts would not handle standard-size cars, "were manageable," Miller says. "There were just too many of them."

Blair blames some of the problems on a geographically distant design firm. Also, the building's systems are highly complex, and there were the sometimes conflicting interests of the computer sciences and electrical engineering departments. Even at prebid, he says, "we had seven addenda on this project."

Thanks to partnering and a disputes review board, there is no pending litigation, says Blair. Everything has been settled, except for a change order request and related insurance claim. But Ellis-Don, partly as a result of this project, has virtually pulled out of the Pacific Northwest market.

Results of a university-sponsored independent report to evaluate capital programs are imminent. But the school has already ordered tighter controls on change orders and will require future projects to be better defined before going to bid.

2

PLANNING AND MANAGEMENT CONCEPTS FOR PROJECT MANAGERS

INTRODUCTION

General Comments

Many observers of architectural, engineering, and facilities management organizations have noted how little planning designers undertake for their own benefit. Design and facilities managers continually preach to their internal and external clients the need to undertake regular planning. However, within their own operations, designers often lack an ongoing planning process, effective management systems, regular evaluation of operations and performance, continual training of their own staff, and a long-range focus. The material that follows is only a brief overview of some planning methods and actions that should be a regular part of your organization's activities. Additionally, a discussion of some key concepts related to effective management and leadership for project managers (PMs) is included.

Planning Concepts

Successful projects require a sound organization-wide project management system and plan. Equally important, individual PMs must prepare specific plans for managing each job under their leadership. There are a variety of types of plans and approaches that PMs need to recognize as they undertake their daily activities.

In design and construction organizations, *long-range planning* helps to conceptualize future goals and needs for programs, facilities, systems, marketing, personnel, and other activities. In comparison to an organization's longer life span, individual projects require *operational* or *short-term planning* to prepare near-term targets in the areas of budgets, staffing, sales, deadlines, equipment, supplies, and for other miscellaneous needs.

Planning for change in the long term requires *strategic planning* to examine the specific businesses, disciplines, markets, regions, and offices that will allow an organization to continue and thrive into the future. Short-term *implementation plans* are often called *action plans* requiring detailed and specific procedures and programs with timetables and accountability in support of strategic plans. *Contingency planning* addresses how best to deal with opportunities, unexpected problems, surprises, or other changes that may arise for your organization or for specific projects.

Short- and long-range planning are common practices in most major businesses. Unfortunately, in many design firms there is a near total absence of these activities. Many designers respond to situations with little or no formal planning to direct and allocate resources, provide for expansion of services and service areas, allow for effective ownership transition, and to organize many similar needs. On the project level, many project managers do not have well-conceived budgets, schedules, work plans, and other important tools. Often, this is a direct result of the lack of long-range planning by design organizations in providing the needed tools, systems, training, and requirements.

Long-range Planning Process

Long-range planning provides a framework that guides choices. A concise definition is: a method to determine the essential thrusts and concepts, and the marshalling of an organization's resources and capabilities toward achieving them. There are four broad elements to this process:

1. **Strategies:** This typically includes the development of broad objectives for accomplishment during a reasonable planning period. This may include the setting of growth objectives, ownership transition in a closely held organization, restructuring into a matrix project management system, or similar objectives that

will require the marshalling of many of the resources of your organization.
2. **Implementation plans:** While strategies describe an organization's broad goals, implementation plans begin to describe how these goals will be achieved. This process provides a general guideline rather than a precise game plan for accomplishing a strategy.
3. **Initial objectives:** After outlining broad goals and general implementation plans, priorities and short-term goals need to be established. Objectives typically cover detailed activities to be undertaken during a series of three- to six-month periods until the stated goal is achieved. Organizations typically hold regular planning meetings to review progress toward achieving initial objectives and to establish new goals as others are completed.
4. **Tactical plans:** These are detailed implementation plans to help achieve initial objectives and eventually, the long-term goals. Tactical plans typically include budgets, schedules, detailed work assignments, and other similar items.

Goal Setting

The setting of goals or strategies is at the heart of the planning process. The three basic approaches to the setting of goals include the following:

1. **Bottom-up decision making:** This approach attempts to achieve the setting of goals or strategies by a democratic agreement of staff and management. This objective, to achieve a consensus by committee meetings without strong leadership and direction, may result in endless discussions with little accomplished.
2. **Pragmatic compromise:** This approach results from senior management offering or imposing a set of general, broad strategies for lower-level managers to implement. These managers are then expected to make proposals or perform activities to achieve these goals. Unfortunately, the results achieved may be based upon what lower-level managers believe senior management will accept. In many situations, this approach discourages real initiative from staff and may, in the end, have little impact.
3. **Incremental or continual decision making:** This method involves developing specific goals or strategies by a mixture of the two other processes. Usually, this planning approach requires a strong leader who has the support and confidence of all managers. Gen-

eral objectives are developed by a planning group with input from all managers. These managers can then make proposals for implementation or work toward achieving goals they participated in setting.

Unfortunately, most long-range planning processes soon stumble. Well-intentioned beginnings often lead nowhere as planning has not been institutionalized in the organization. Managers and staff become deeply involved in the day-to-day activities of their job and little focus is given to the long-term needs of an organization. Only when failure to address systemic problems creates a crisis in an organization are these planning needs given attention. By that point, it may be too late to correct the problem(s). To avoid crisis management, senior managers and project managers must commit to a continual planning process and seek continuous improvement in both firm and project operations.

MANAGEMENT CONCEPTS

Decision Making

In design and owner organizations, PMs are unique individuals. They possess a special set of skills different from all others on project teams. Designers are concerned with the form of a building, road, bridge, or other facility. Technical staff is concerned with function—they need to ensure that the lights work, that the heating and cooling systems operate properly, that the structure is sound, that the roof doesn't leak, that the building is safe for people to be in, and numerous other details. Owners are concerned with the cost of a facility, whether or not it meets their functional needs, how difficult it is to maintain (life cycle costing), and so on. PMs need to be aware of all of these concerns along with many of their own, including meeting a design and construction budget, schedule, and program. They must apply sound planning and management techniques if they are to be successful in their work.

Management theory requires that PMs follow a disciplined approach to their decision-making efforts. Normally, this requires five steps.

1. **Define the problem and needs:** In an owner/client organization, this requires extensive effort to achieve a consensus on proj-

ect goals, objectives, problems, solutions, and approaches. Some owner organizations have large and experienced facilities staffs (e.g., the U.S. federal government), who devote a great deal of effort to define or resolve many of these issues before design consultants are hired. Other owners hire specialized consultants to define needs and goals and then retain architects or engineers. A third option, now less common than in the past, is when owners/clients hire architects or engineers to help define needs, goals, and objectives. On the project level, it is important to include PMs in the effort of defining goals, objectives, and needs for their own project. The PM is ultimately accountable for project profitability and performance and must have a major role in this process.

2. **Analyze project needs and requirements:** In owner organizations, the facilities staff is often brought into the definition-of-need process after many key decisions have already been made. Unfortunately, this means that the people who know the most about the design and construction process are the last to be consulted. It is not uncommon for client needs to become a moving target as the project program continually changes. This occurs because internal user needs were never fully defined in the first place or because internal politics create a competitive environment where a group or a manager places their own self-interest first. The design firm PM often ends up attempting to reconcile the needs and demands of the various groups within the owner's organization. Without this reconciliation, it becomes extremely difficult for designers and technical staff to focus on the project needs and requirements and to move toward solution of the client's problems. In design firms, the PMs also must analyze the project and determine their own fees, schedules, team members, and approaches.

3. **Provide alternatives for solutions:** PMs are the key communications link in an organization involved in the design and construction process. Externally, their role is to represent their own firm to the other members of the project team and to communicate their concerns, problems, and solutions. They also need to encourage a smooth working relationship with all others involved in the project. In their own firm, they must encourage, motivate, and push their staff to perform in the best interest of the project and their firm. They must attempt to reconcile the differing priorities

of designers and technical staff and seek compromise when necessary. A PM's own design ability and technical knowledge must be sufficient to come up with solutions and alternatives when others are unable to. When various options and choices exist, it is important that project managers avoid imposing their own solution to a problem. It is vital that those who have the assignment as a project's designer, technician, or other role be given full opportunity to undertake their assignment. PMs who jump into each and every issue that arises and usurp the responsibility and authority of others will quickly find themselves overwhelmed with detail and unable to perform in their primary roles as leaders of the project teams. Leadership does not mean doing everything yourself. PMs who forget this important point will likely fail.

4. **Compare, contrast, evaluate, and choose:** Decisions are required on all design and construction projects. Committees and experts must have the opportunity to weigh in with opinions and advice. Eventually, someone must be responsible for a final decision and this is an important part of the PM's job. Indecisive people do not make successful PMs. At the same time, it is important to listen to others. Seek out advice and carefully evaluate options. Often, all options will have drawbacks and limitations. A PM must be willing to choose what is believed to be the best from among a list of uncertain choices. This situation was best highlighted by Percy Barnevik, former chairman and ex-CEO of Asea Brown Boveri, one of the largest electrical engineering companies in the world, and quoted in *Information Strategy* magazine in December 1996. He stated, "I tell my people that if we make 100 decisions and 70 turn out to be right, that's good enough. I'd rather be roughly right and fast than exactly right and slow…the costs of delay are vastly greater than the costs of an occasional mistake."

5. **Implement:** Once decisions are made, it is important to communicate them to all members of the project team, both internal and external to your own organization. It is crucial to have an effective system to communicate decisions and information to all parties for implementation. Design and construction are complex undertakings involving vast numbers of details and decisions. Err on the side of sharing too much rather than too little information. A very common complaint in design is, "no one told me!" Changes made by an architect are often not conveyed to the consulting engineers and frequently the reverse is true. Unfortunately, contractors and

subcontractors receive drawings with differing or contradictory information from various design team members who are working from different sets of data. The evolving concept of Building Information Modeling (BIM) is attempting to solve this problem, but BIM is not yet a mature approach and the interoperability of computer software remains an obstacle.

Roles of a Project Manager

A successful PM must have excellent communication skills. Project decision making requires leaders who can communicate information to others to allow for effective implementation. One commentator noted not long ago that as the technology to communicate has improved, our ability to communicate has become more ineffectual. Fundamental to a PM's success is an effective decision-making system. Lines of communication must be clear. A PM's authority must be commensurate with his or her responsibility. Needed information must be accessible and timely. Individuals must be clear about their role within an organization.

There is a great deal of management literature on the appropriate role for a manager. What follows is a brief summary of thoughts from an expert on the subject. This material is drawn from *Getting There by Design: An Architect's Guide to Design and Project Management* (Allinson, 1997). Kenneth Allinson extensively refers to Henry Mintzberg, who in 1975 conducted a study of "what managers actually do as opposed to what theorists suggest they do." As summarized below, Mintzberg noted that managers have ten roles to fulfill:

1. The manager is a symbolic figurehead.
2. The manager is a leader offering guidance and motivation. This is among the most significant of a manager's roles.
3. The manager plays a liaison role, dealing with a web of relationships.
4. The manager must monitor what is going on in areas such as internal operations, external events, analyses, ideas, trends, and pressures.
5. The manager must play the role of disseminator.
6. The manager is a spokesperson to the outside world.
7. The manager is an entrepreneur acting as the initiator and designer of controlled change within the organization. This role particularly concerns projects seeking to utilize opportunities to

solve problems over which the manager acts as delegator or supervisor. As Mintzberg notes, the inventory of projects changes continually as new ones are added, old ones reach completion, and others wait in storage until the manager can begin work on them.
8. The manager is a disturbance handler, reacting quickly and dealing with conflicts between subordinates, difficulties between organizations, or resource losses.
9. The manager is a resource allocator which is a role at the heart of strategy-making. Time must be scheduled, work must be programmed, and actions must be authorized. Decisions are regularly modeled in the manager's head.
10. The manager must act as a negotiator, responsible for representing the organization.

These ten roles for the manager were grouped by Henry Mintzberg into three areas. Roles 1-3 he called *activities concerned with interpersonal relationships*. Roles 4-6 he called *informational roles*, and the remaining four were called *decisional roles*. Mintzberg emphasizes that managers undertaking these roles rely strongly on intuitional judgment as opposed to the rational-deductive approach taught in most business schools.

SUMMARY

The material in this chapter is not intended to be a substitute for an intensive study of business and management practices in the design and construction industry. The goal was to cover a few significant concepts of importance to PMs. This material focuses on planning, goal-setting, creating a structured approach to decision making, and understanding the appropriate role for a project manager. Henry Mintzberg is correct—intuition is of vital importance to your success as a PM. Trust your judgment and experience.

This book has free material available for download from the Web Added Value™ resource center at www.jrosspub.com

3

THE PROJECT MANAGER

WHO IS A PROJECT MANAGER?

Survey Findings

A survey conducted by the Association for Project Managers (APM) shows that design firms and owner/client organizations have vastly different definitions of the role of project managers (PMs). Most, however, claim that many or all of their jobs are managed by PMs. More than 80 percent indicated that their projects are managed by an individual with this title. About 58 percent of the firms surveyed had five PMs or less. The median (midpoint) number of PMs when including all firms in the survey was five. Finding and keeping capable PMs remains a vexing problem for many firms. Nearly two-thirds (66.1 percent) of the respondents reported difficulty in hiring and retaining managers. Turnover in the PM position is another significant problem, with nearly 48 percent of firms suffering some loss in the ranks of these key individuals in a one-year period. Figure 3.1 highlights this finding.

In many design firms, principals often assume project management responsibilities. For small practices this is necessary because of the lack of sufficient staff to serve as PMs. In other cases, it is the principal's preference to undertake project management in addition to his or her other firm management and marketing responsibilities. PMs tend to be among the most highly experienced individuals in design firms and facilities organizations in design and construction activities. The mean number of

NUMBER	FREQUENCY	PERCENT
0 or blank	34	52.31
1	14	21.54
2	6	9.23
3 or more	11	16.92
Total	65	100.00

Note: missing = 2

Figure 3.1 Number of project managers leaving

years of professional experience for PMs was 12 and the median was 11.5. The range was from 3 to 26 years.

Characteristics of Project Managers

Few university technical curriculums provide even rudimentary project management training. Necessary skills are typically learned on the job, through seminars, or from self-initiative. Facilities managers and owner PMs are frequently placed in their position with little formal training or guidance. Many designers have minimal ability or interest in the business side of architecture or engineering. Often, those who show even the slightest inclination toward management are pushed toward becoming PMs.

Who should be a PM? How are they created? What characteristics are important? Few would dispute the fact that the best technicians often make the poorest PMs. The reason for this is obvious. Most individuals with a proven technical ability tend to focus strongly on the aspects of a project with greatest interest to them. With few exceptions, this focus works to the detriment of the broader project needs a manager must address.

What capabilities should a PM possess?

1. **Strong organizational skills**: The successful PM must be able to organize a project, the team, and address the many details that arise. This employee must be strong at organizing staff schedules and be able to handle more than one major project if necessary.
2. **Generalization:** While a manager may have an interest in a particular project area, he or she must be familiar with all aspects of the project. However, PMs do not need to know all of the technical

details for themselves. Being effective at delegation is very important. To be effective as a PM, one must have a strong ability to examine the broad scope of a project without becoming bogged down with the details.
3. **Ability to monitor the project:** A PM must be able to monitor project status and display a willingness to ask for assistance when the situation warrants. Effective communications between members of the project team is vital.
4. **Communication:** PMs must have good communications skills for both public speaking and writing. They must be able to communicate to both individuals and groups as marketers and as managers of the project team. In addition, they must be good listeners.
5. **Experience:** Successful PMs must have broad experience in a variety of project types. They must have strong skills and experience in project budgeting, negotiating, marketing, and estimating. Their own database of previous project experiences is extremely useful.
6. **Leadership ability:** The strong PM must be a leader; an individual who can direct and motivate the project team. He or she should have demonstrated leadership ability prior to becoming a PM.
7. **Ability to make decisions:** A PM is a decision maker. The ability to make decisions and to carry them through is vital. In addition, the manager must be able to admit a mistake, and must be able to say no to a client or staff member when necessary.

Communication Skills

The key function of a PM is to communicate. He or she serves as the primary link between members of the project team. Each design consultant, contractor, and client should be represented by a PM who is able to communicate their needs, questions, and status to other team members. To accomplish this function, PMs must be skilled communicators. Public speaking skills must be learned and polished through extensive practice or with formal training. Writing skills must be developed to a high degree. PMs should attend university or community college writing courses to learn fundamentals, technical writing, and persuasive writing. Firms should retain, on a full- or part-time basis, an internal staff member to review and constructively criticize written materials and to train all staff in effective written communications.

SUMMARY

In general, most of the qualities and characteristics of the successful project manager revolve around his or her ability to work well with people, rather than technical skills. Certainly, the PM must have basic technical skills, but overemphasis on these by senior management will not necessarily result in a good project manager. All project managers require regular training to improve skills and to learn new ones. This training must become a regular part of creating effective project managers.

CASE STUDY: OHIO OPERATING POWER SYSTEM

The Ohio Operating Power System (OOPS) is the primary electric power utility company in the eastern part of the state of Ohio and the southwestern part of Pennsylvania. Approximately one-third of the utility's generating capacity comes from its nuclear power plants. The company is best known for the near meltdown of one of its nuclear plants in the late 1970s. This plant, historically known as the *Butler Ohio Operating Mill* (BOOM) was named after the early 19th-century grain mill located on the same site.

OOPS has been cited repeatedly by the U.S. Nuclear Regulatory Commission (NRC) for sloppy maintenance and poor management practices. In evaluating the problems of the BOOM plant, the NRC stated: "OOPS continually failed to follow its own rules and regulations, failed to properly train its operators, and often left the operation of its nuclear plants, coal and oil fired plants, and other facilities in the hands of unqualified staff and managers." Further, the NRC noted: "The Company paid no regard to prudent planning, evaluation, fiscal, and reporting procedures."

In response to the NRC criticisms, the poor public relations of the BOOM plant crisis, and state regulator pressure, OOPS hired a series of specialized consultants to evaluate all operations and to recommend changes. Concurrent with the NRC investigation, the entire U.S. utility industry faced deregulation. Until deregulation, public utilities operated with little fiscal oversight. State regulators were either unable or unwilling to effectively monitor the expenditures of utilities. Most public utilities were free spenders to the point of total irresponsibility. Indeed,

Jonathan Kleinman, head of the Public Service Commission in Pennsylvania commented: "OOPS and other utilities spend money in a manner consistent with an inebriated mariner on leave."

Under deregulation, large commercial consumers of electricity were now free to purchase their power on the open market. As all electricity is essentially the same, price became the prevailing factor in its purchase. Although OOPS and other utilities had several years notice of the impending deregulation, they often failed to adequately prepare for this new economic environment.

Free spending and poor planning at OOPS was not limited to its generating plants. The general turmoil in the company led to frequent staff and managerial turnover and regular *reorganizations*. The only consistent staff members were the plant managers who oversaw the generating plants and other OOPS buildings such as offices, warehouses, maintenance facilities, and the like. Typically, these individuals were high school educated, *blue collar* types who focused on mundane day-to-day issues. None had management training, financial acumen, or strong communications skills.

The utility maintained an extensive office complex in Zanesville, Ohio. It was common practice at the utility to reassign department managers on an annual basis. Each new incoming manager had his or her own concept of the operation of the department. Often, this required a physical redesign and reconstruction of office space. This process was nearly continuous and ensured the regular presence of construction personnel—all of whom were OOPS employees. Outside contractors or subcontractors were rarely used. There were no architects, engineers, or other trained design professionals on staff. Except in the most complex situations, drawings of the planned new space were never prepared. Spaces were built out and new department managers would often request significant changes after completion. Less than a year later, the process would begin again. Compounding the situation was a budgeting process that was based on a *spend it or lose it* program. No benefit accrued to a manager who kept spending under budget.

When deregulation arrived, it spread fear at OOPS. In the past, if wild spending resulted in the need for more money, the utility went to the state public service commissions (PSCs) for a rate increase. A common strategy was to ask for more than was really needed to allow the PSC room to cut the request as a symbol of their *protecting the public*. It was a commonly played game. The PSC never turned down a rate increase. The free market never impacted a public utility.

Deregulation was expected to require a significant change. For the first time, budgets were prepared, long-term plans outlined, and work documented. In the OOPS construction operation, it was determined that PMs needed to be identified. The goal was to place these individuals in charge of the design and construction of all facilities. They were to be the in-house manager representing OOPS interests when outside design firms and contractors were hired. They were also to manage the process when internal OOPS construction staff was used.

As an insular culture, senior OOPS managers sought to elevate internal staff to the positions of PMs. Unfortunately, there were no individuals qualified for the positions. Rather than hire outsiders, it was decided to select a group of plant managers and broaden their responsibilities. Without exception, the group members selected were all in their 50s and 60s and nearly all had been lifetime OOPS employees. Most were looking forward to retirement and were not at all enthusiastic about their new responsibilities. The modest pay increases they were granted did little to encourage them.

A basic training program was developed to assist these new PMs. The initial two-day session brought in an expert in construction industry project management to provide an overview of the subject. The consultant expected little participation on the part of these novice PMs and he was not surprised when they greeted his presentation with blank stares. What he was not prepared for was the outright hostility expressed by these individuals. They made it clear that they didn't want to be there and had no intention of exerting themselves in their new role.

It was no surprise when the training program fell apart and the new PMs continued to function as they always had. Little in the OOPS construction operation ever changed.

THE PROJECT MANAGER

Areas Requiring Attention

One of the most significant changes in design firm and owner/client organizations in recent years is a renewed focus on project management. Gone are the days when any organization could get by with poor project management practices. Shorter time frames and tighter budgets are forcing firm managers to make significant changes in their operations. The following areas require attention:

1. **Organizational systems:** While many organizations have established project management systems, unfortunately, some have a weak version of it. While PMs may have been appointed, senior managers are often slow to give up control over projects. As a result, PMs have titles and responsibilities, but lack effective authority to make decisions stick. For firms to be successful at project management, a true delegation of responsibility and authority must occur. Regular education on the role of project management in firms is also required. Senior managers must understand the appropriate organization of a strong project management system and the role of the individual PM.
2. **Recruiting:** Project management requires a unique set of skills often lacking in technically trained individuals. Many firms lack a clear idea of the role of the PM, and as a result, they are uncertain as to the needed skills. As noted earlier, some firms select PMs based on technical expertise, not on needed leadership and management skills. To be effective in recruiting needed staff, many well-managed companies establish career tracks. This may include a design, technical, or management track. Position descriptions are clearly written, organizational structure is established, and individuals are recruited or promoted to fill a defined need.
3. **Training:** Few firms have adequate training programs. This lack of training strongly impacts PMs who require the broadest range of skills of any firm member. A good training program will have several goals:
   - Training staff in the firm's methods of operation
   - Providing replacements/support for existing staff
   - Improving job performance and productivity

   For many organizations initially developing a project management system, an individual experienced in project management should be hired to guide the effort. Often, this individual brings an understanding of how the system should function. He or she may bring forms, checklists, and procedures that can speed up the implementation process. This manager often provides initial training to other PMs.
4. **Productivity:** The years ahead will continue to require enhanced focus on how to improve productivity. Specialization can affect firm productivity and may require PMs who are experts in a par-

ticular type of work. Providing the necessary tools and information to these key individuals will allow them to improve job performance and will enhance productivity.

PROJECT MANAGER RESPONSIBILITIES

General Comments

There are as many definitions of a PM's responsibilities as there are firms. In some organizations, a PM is anyone who tells someone else what to do. In one extreme case, a 50-person architectural practice claimed that 20 people were PMs. There is no viable rule of thumb for the number of PMs needed for a certain number of jobs; however, where a full-charge PM runs the job with the proper tools and staff, a large number of projects can be managed by one individual.

Specific Responsibilities

The following is a listing of some of the major responsibilities of PMs. Clearly, this is not a complete list, so firms developing a PM position description should seek additional resources. Many are specific to design firms while others apply to both owner/facilities managers as well as designers.

1. **Marketing and continued contact:** Most experienced clients want to meet with and discuss their project with the individual actually responsible for the job. Initial contact and discussions should be held with the design firm marketing staff or principal; however, the PM should be brought into the process as early as possible. The design firm PM should also function as a source of continuing contact with clients on completed projects. This is to ensure the smooth functioning of the facility and to be aware of further services required by clients.
2. **Proposal preparation/fee determination and negotiation:** One of the most important responsibilities of design firm PMs is the preparation of a proposed scope of services and the corresponding fee. As the individual responsible for meeting the scope and fee, the PM must have a major role in their preparation. Scopes and fees imposed on project managers by senior management will not allow for full accountability. With less experienced managers, a

thorough review process is essential. While final contract signing must be done by an officer of the design firm, the PM should lead the fee and scope negotiating process. This allows for a full understanding of these items in the event that adjustments in scope or fees are required.
3. **Staff planning and assembly of the project team:** The PM is closest to the project and the required staff needs and schedule. He or she must provide regular input to senior management to allow planning for overall firm staffing needs. As the project begins, the PM must suggest names of specific individuals he or she would like to work on the job. This is based upon their knowledge of the needs of the project and the specific skills of staff members. Obviously, the suggested list must be adjusted based on staffing needs of other projects and on the availability of various individuals.
4. **Managing the project:** The PM is responsible for supervising all phases of the project, including time charges, meeting budgets (both fee and construction), and handling the many other details of guiding a project. While most PMs should not be making technical or design decisions, in some cases this becomes necessary to ensure compliance with the program.
5. **Quality management:** Quality management is a shared responsibility. However, PMs must ensure that quality reviews are budgeted into the project fee and occur at appropriate times. In some organizations, where the PMs have the necessary technical competence, they may actually review drawings themselves, although this is generally not an appropriate use of their time. It is the responsibility of senior management to develop a quality assurance program—the technical staff and the PMs must ensure that the program is instituted.
6. **Team relations:** It is the PM who can best communicate with the various internal and external team members. There must be a regular process of meetings, e-mail distribution, telephone calls, correspondence, etc.
7. **Project status reporting:** PMs must not only prepare the original project budgets, they must also support the reporting system by providing accurate updated budgets, contract data, and percentages of completion (design firms) on a timely basis. Information, including revised budgets and maximums on change orders,

needs to be current. Project status reports must be monitored regularly.
8. **Billing and collection:** Although in most design firms invoices are prepared by an accounting office, the PM should review and approve all invoices prior to issuance to clients. PMs should not be responsible for collecting outstanding balances, but considering their close relationship with clients, they may be able to expedite collections. Additionally, PMs must understand a client's billing requirements (supporting documentation, schedule, etc.) and supply this information to their accounting offices.

Design Firm Project Managers and Marketing

PMs can and should have a major role in a design firm's marketing effort. Unfortunately, many engineering and architectural firms fail to properly integrate their PMs into the marketing team. This failure can seriously affect the marketing success rate and future project profitability. Well-managed design organizations train their PMs to be important players in the marketing phase of projects. To be effective in the marketing effort, PMs should:

1. **Be assigned to a potential project at an early stage:** Once it has been determined that a lead is materializing into a tangible project opportunity, a PM should be assigned. He or she should participate in subsequent marketing activities and later manage the project if the effort is successful.
2. **Serve as a primary contact person for the future client:** This role is shared with the senior marketer. The PM must be the client's point of contact on technical and scope issues while the senior marketer/principal is the liaison on other issues. Clearly, the PM and marketers must meet regularly and maintain effective communications.
3. **Provide technical input:** With the help of the technical staff, if necessary, the PM should provide technical input to both the marketing staff and the future client. This role may include assistance in the preparation of a preliminary project program or the suggesting of alternatives that may have lower initial or life cycle cost, allow for easier expansion, and so on. As a result of this effort, firm members may demonstrate their experience,

knowledge, and strong concern for the future client's needs and budget.
4. **Participate in the presentation process:** Most experienced clients want to meet the individual who will be responsible for managing and performing the work. The presentation process is an ideal time to strengthen the relationship between the design firm PM and the potential client. The presentation allows the PM to provide specific information as to how the project will be managed. The PM should have an active role in preparing materials and strategies for presentations.
5. **Develop a detailed project scope:** Clearly, the PM has far more experience than most other marketers in the development of the project scope. With early involvement in the marketing effort, the PM should be very familiar with the future client's stated and implied needs, budget and operational concerns, internal decision-making process, and staffing. PMs are uniquely prepared to outline a proposed scope of services and to evaluate where adjustments in this scope can occur. In addition, it is vital that the individual who will eventually be responsible for delivering a scope of services to the client be involved in the preparation of that scope.
6. **Develop a detailed project budget:** The PM must, in the development of the project scope, be aware of the costs required to complete the proposed scope of work. He or she must assemble a detailed project budget and outline areas where the budget can be altered by a change in scope, or by negotiation. Without this total understanding of the proposed project budget and scope, the negotiation process will be needlessly complicated. Only when the PM has actually prepared the project budget can he or she be held accountable for it.
7. **Participate in the negotiation process:** With a full understanding of the proposed scope and budget, the PM is invaluable during final negotiations with the client. Any adjustments that need to be made should be based upon the PM's evaluation of the program and upon the ability of the firm to make an adequate fee and profit for its work. No commitments should be made to the client without the PM's understanding and agreement.
8. **Provide input into future staffing and other resource requirements:** A successful marketing effort commits the firm to supply-

ing staff and other resources to the client's project. The PM must fully understand these needs and communicate with other managers to ensure resource availability. This is an extension of the marketing effort since failure to plan for these needs will impact the service provided to the client.
9. **Provide the design firm with continued contact with the client:** In all firms, upon completion of a project, personnel and resources focus on other projects. As a result, many past clients are inadvertently neglected and little follow-up is conducted on project and building performance. When the client is in need of future professional help, much, if not all, contact with the design firm may have been lost. To prevent this, the project manager must contact the client regularly. The PM must also be aware of opportunities to suggest changes or improvements in the existing facility.

In addition, PMs should be alert to opportunities to inform a client about the firm's other services. Often, clients have a perception based upon the services the design firm is currently providing to them. Clients may be unaware of a design firm's additional capabilities and thus not consider the firm when awarding other projects. Every staff member has a responsibility to assist in the marketing effort. PMs, however, are in a unique position to provide job leads, make new contacts, and become involved in community and professional organizations. It is through these efforts that the PM can also contribute to improving the success of the firm's marketing program.

CARING FOR AND TRAINING YOUR PROJECT MANAGERS

Finding Project Managers

A continuing source of difficulty for many organizations is finding, recruiting, and keeping capable PMs. As noted previously, few engineering, architectural, or facilities management programs teach management skills to any degree. As a result, there is a significant shortage of skilled PMs. As increasing numbers of firms recognize the value of project management, the competition for available talent is nearing crisis proportions. In large cities, high job mobility creates the opportunity to recruit PMs from other firms. In many smaller cities, however, the total architectural, engineering, and facilities management community may only number in the hundreds. As a result, experienced managers may be unavailable or cannot be recruited from other local firms. For many firms, there are three basic techniques in obtaining the required talent:

1. **Recruit from outside your organization:** This method is often the fastest approach to building your project management staff. Recruiting from other local firms (particularly in smaller communities) may create animosity on the part of your peers, and may also eliminate any hesitancy other firms have about raiding your staff. In addition, the local design community may be somewhat inbred and firms may simply be exchanging each other's weaknesses. If the local pool of talent is thin, recruiting from other, usually larger cities may be the solution. Unfortunately, attracting staff to smaller communities may not be possible when seeking highly paid, experienced PMs. Offering competitive salaries, fringe benefits, and ownership (or potential) has been used with varying success.
2. **Train your own project managers:** In some communities, the only significant source of PMs may be in a firm's own staff. Some firms are reluctant to make a major investment in training their staff for fear of incurring the expense only to lose these people to competing firms after a few years. Clearly, a certain percentage of your staff may leave for various reasons. With sufficient incentive (e.g., salary, bonus, ownership, profit sharing), many capable staff will remain to help the firm prosper. These individuals will have made the cost of training worthwhile. This training process requires constant budgeting of time and resources for seminars, courses, and publications. Some firms recruit prospects directly from colleges and universities to obtain the most capable talent. They then educate these individuals into PMs compatible with their organization's philosophy. In large cities, successful firms with experienced teams of PMs also seek younger talent and bring them in as assistant PMs to fill needed slots. In many locations, it is not unusual to find a large percentage of design professionals who have worked for one or two local firms at one time. Many of these firms are noted for their training programs.
3. **Recruit an experienced project manager as the mainstay of your staff:** For many firms not experienced with effective project management, it is often wise to recruit one knowledgeable manager

as the center of your system. This individual should help establish the project management program, recruit and train younger staff, and serve as a technical and managerial resource. In some organizations, it may not be necessary to recruit an experienced manager, since a senior manager may wish to begin an intensive self-education program to acquire the necessary skills.

Keeping Your Managers

Finding and training your PMs is only the first step. Keeping your hard-won managers is just as important. It is the responsibility of senior management to provide for the psychological and financial well-being of these individuals. The obvious incentives of competitive salaries, bonuses, profit sharing, and a fringe benefit package are most important. As noted earlier, PMs must have the authority that matches their level of responsibility in the firm. Second-guessing and countermanding their decisions will quickly destroy your system. As a result, many of your managers may become interested in opportunities with other organizations.

How Many Project Managers Do You Need?

A frequently asked question is how many PMs does an organization need? Unfortunately, there is no fixed answer—for a number of reasons.

1. **The experience level of your project manager:** Generally, more experienced PMs should be able to handle more projects. They should also be capable of managing projects of greater complexity than managers with less experience. In many cases, firms employing less experienced managers may require a greater number of individuals to handle the workload.
2. **The experience level of your technical staff:** Staff who are experienced in a particular project type can make the PM's job a great deal easier. Their familiarity with the workings of the firm's project management system is also important.
3. **The quality of your support and management information systems:** A quality system will impact the ability of PMs to perform. Effective project management requires a wide array of tools and systems. When managers lack all or part of these systems, more time is required to complete their project responsibilities. This may mean the organization will need a greater number of PMs.

4. **The complexity of the specific project:** When a manager handles a complex project, it will affect his or her ability to manage additional projects. Managing complicated medical or manufacturing facilities will require greater effort and attention than basic speculative office buildings or retail spaces. Increased time demands may be placed on a PM to administer a complex project.
5. **The geographic location of projects:** The manager's work time will be impacted by the location of the project. Projects located in distant or remote locations will involve more travel time for meetings, job-site visits, and for general coordination. While technology can be a great help in managing projects such as these, they will not be a complete substitute for personal contact.
6. **Experienced and sophisticated clients:** For design firms, experienced and sophisticated clients can ease the burden on PM. The smoothly functioning relationship between the design PM and an experienced client can allow a PM to handle additional projects. Occasionally, the sophisticated client can actually require more of a PM's time by demanding additional services or attention. However, this may be preferred to the hand-holding required with an inexperienced client.
7. **Staff experience level:** A major factor in determining the number of required PMs is the general staff experience level with the current projects of the firm. When a one-of-a-kind, rarely handled project type is encountered, the learning curve is higher for all. Increased levels of research will be required, as will increased levels of team interaction. This will usually necessitate greater time commitments for the PM.
8. **The current number of projects under contract is an important factor:** Most experienced PMs can handle several projects at one time. An increased number of jobs in a design office will result in a higher workload per PM or may require the hiring or training of additional PMs.
9. **The mix of projects normally handled by the design firm:** The types of projects that are in progress can affect the number of PMs. For example, an office with a large number of small projects will likely need more PMs than one with a few large jobs. Since most projects need at least a minimum level of service, despite their size, more projects means a higher workload.

10. **The timing of your projects:** Projects with similar time schedules can put extraordinary demands on PMs. Most design firms strive to have PMs handling projects in differing phases to avoid crunches. Typically, one project may be in a start-up phase, another heavily into design or construction documents, and a third may be in construction administration. Unfortunately, delays and changing programs may alter this desired scheme. At those times, the pressure on PMs may be extreme.

With no easy way to determine the required number of PMs, how do firms cope? Well-managed organizations are always training younger staff in the principles and applications of project management. Junior staff can assist experienced PMs as a method of learning required skills. Over time, they may be given the opportunity to manage their own smaller, less complicated projects. In this way, individuals are available to meet the needs of an ever-changing workload. In smaller organizations with limited staff, most technical employees should be given basic project management training. It is vital that the required tools and systems are in place to assist in managing projects.

CASE STUDY: AB&C TELECOMMUNICATIONS

AB&C Telecommunications is the largest company of its type in the United States. The present firm is a result of various mergers and acquisitions made throughout the 20th century. Until the 1980s, most of AB&C's growth was internally generated, as the firm had a near monopoly on telecommunications services in the United States. Innovation and a competitive spirit were lacking in the firm. With the breakup of the monopoly by court order in the 1970s, the company was faced with numerous new and aggressive competitors. Many of these competitors captured market share from the lethargic remnant of the old monopoly. Even some of the pieces of AB&C spun off by the court order became serious threats.

Struggling to right itself, AB&C was able to maintain a substantial market share by clever anti-competitive devices. Early in the new competitive environment, the firm convinced most state regulators to assign new customers to AB&C as the default provider of telecommunications services. Consumers had to actively elect to choose another provider. Most did not make this effort. AB&C also was successful in the short term in charging other telecommunications providers for services such as the use of switching equipment and other existing infrastructure. They claimed that these fees were in compensation for the expense AB&C had incurred for creating the infrastructure others were now also using. This argument ignored the fact that past AB&C customers had actually paid for this equipment as charges on their monthly telephone bills. The net result was that new providers were faced with passing on these charges to their customers resulting in higher bills. Eventually, however, most of these anti-competitive tactics failed and the playing field was leveled.

AB&C now needs to compete based on service, innovation, and price. Competition has become cutthroat and new players are regularly entering the marketplace. Technological innovation is blurring the lines between equipment capability and providers. Telephone calls can be placed using computers. Telephones, particularly cellular phones, now have fantastically broad capabilities. Any provider with sufficient bandwidth availability can offer telecommunications services. Big, old AB&C hasn't been agile enough for many.

The company regularly reorganizes. A recent approach divided the firm into two basic units, the network and related systems, and everything else including all non-network infrastructure such as buildings, warehouses, etc. Design and construction (D&C) operations were likewise divided. Regions were established and a head of D&C appointed in each location. Duplicate staffs for network and non-network D&C were established and were composed almost exclusively of PMs who functioned as owner's representatives. The clients of these PMs were internal AB&C departments and locations.

Determination of need and project programming were largely handled by the end user—the internal AB&C client. D&G budgets, schedules, and other requirements were also developed by the internal client, rarely with the D&C's input. The standard operating procedure allowed D&C to review the materials prepared by the user and comment back to them. Often, D&C objections and concerns were ignored or minimized. Despite this, the D&C PM was expected to meet the internal client's needs. This resulted in a classic situation where the AB&C PM had the responsibility for the project, but none of the needed authority over planning and resources.

In recent years, in an effort to cut costs further, AB&C has turned to outsourcing the PM's work. In many cases, AB&C fired the D&C PMs and then hired them back as *consultants*. While the base hourly pay rate for the PMs rose, AB&C eliminated expensive fringe benefits such as health care and retirement programs. This tactic also gave the firm great flexibility in staffing. If work declined, AB&C simply hired fewer consulting PMs or greatly reduced their hours. AB&C even helped a former employee establish a project management consulting firm to organize and manage the former D&C PMs. This firm handles billing, payroll, health insurance, etc. While many of the new *consultants* enjoyed the freedom and flexibility of their new role, others realized that they were now highly dispensable, with few job or work guarantees.

Today, AB&C remains in a cost-cutting mode. D&C work is only approved when it can be justified in great detail. Most D&C work is still managed by independent contractor *consultant* PMs. On paper, the D&C organizational structure remains largely the same as that established several years ago. Now, however, it is only a shell organization consisting of a few managers without any actual staff.

Many client/owner organizations view design and construction staff as extraneous to the core *mission* of their business. As a result, this operation becomes an easy target for cost cutting. Some firms once employed large numbers of their own architects and engineers to handle facility design and construction needs. Few companies or institutions retain these today. Most have long ago out-sourced this function. Some are even following the AB&C model and eliminating in-house project and facilities managers. They are turning this job over to contract staff or outside firms such as program or project management consultants.

REWARDING PROJECT MANAGERS

Senior managers in many organizations find it difficult to determine appropriate rewards for PMs. As key employees of the firm, PMs deserve a high level of base compensation. In most parts of the country, full charge PM salaries begin at no less than the mid-$60,000s—and can range upward of $150,000 (or more) in major cities and in large firms. Salaries for PMs working on complex international projects can be substantial and they may receive generous living allowances.

Beyond base salary, what are some cautions to consider when rewarding PMs? Smart firm managers generally try to assign their best PMs to the most difficult projects. This may be the most technically complicated job—one that has a tight time schedule or a very restrictive fee budget. Basing a PM's rewards in a design firm upon the project's eventual profit level will be extremely unfair to the good performer assigned to a tough job.

Despite this caution, some design firms have attempted to base both the PM's reward and that of the entire project team on the final profit level of the job. During the fee budgeting process, senior management and the PM set a target profit level for the project. Any profits beyond that level are divided among the members of the project team, with the manager receiving as much as half. As a result, short-term gains may be achieved at the expense of other projects and the entire firm. Obviously, if a PM stands to gain financially from a project making a substantial profit, they will likely put forth a significant effort on that particular job. A project with a lower profit potential due to size, complexity, or other factors may receive less attention. Some firms provide a down-side risk to PMs. They are penalized for projects that lose money and this is deducted from the profitable ones. However, if losses exceed profits, the system nets out to zero and no further penalty is incurred.

PMs typically do not have control over resources such as the assignment of team members, schedules, and often, the selection of consultants. It is these resources that can make the difference between a profit and a loss. Sharing project profit with PMs and team members can encourage short cuts, poor client service, and destructive internal competition.

Positive Rewards

Firms must provide an atmosphere conducive to positive self-esteem. As noted earlier, PMs must not only have a high level of responsibility, they must also have a corresponding level of authority. Senior management must trust the judgment of their PMs and publicly back their decisions, even if private follow-up is required. PMs must also have significant input into decisions for which they will later be held accountable. For example, a project fee budget should be prepared by the PM and reviewed by senior management, not the reverse. This process rewards and encourages the PM by providing a positive environment of trust and confidence.

A spot bonus program or other reward is sometimes implemented to reward good performers. While the amount of money is not significant, the symbolism is very important. Salary increases and promotions are obvious rewards for good long-term performance. The use of increased fringe benefits can also be a valuable reward. This might include the providing of a company car or similar reward. Firms should also have a 401k or similar program in place to reward all staff members for superior year-long performance.

Training Project Managers

Few members of your staff are of greater importance to your success than PMs. Their pivotal role requires them to possess a unique set of skills. Unfortunately, many PMs are forced to learn on the job. The benefits to them and the firm of a formal training program are great. What should a PM training program include? There are three broad areas that should be covered: communication skills, interpersonal skills, and technical management skills (see Figure 3.2 for a suggested list of training topics for PMs).

1. **Communication skills:** PMs need to possess a wide range of communication skills. In particular, negotiating, effective writing, and public speaking are vital. Their involvement in negotiating contracts and with other members of the project team makes this an obvious area of focus. Many design and construction staff are poor writers. Much of their writing suffers from wordiness, improper punctuation, capitalization, run-on sentences, and a long list of other grammatical faults. This inability to write effectively and properly reflects poorly on your organization.

    Public speaking opportunities for PMs include community groups, social organizations, client meetings, project team meetings, and marketing situations. Public speaking is high on most individuals' lists of major fears. This fear is often only overcome through practice. While some PMs practice presentations before project teams or family members, others seek more directed and instructive environments offered by groups such as Toastmasters.

2. **Interpersonal skills:** PMs are people managers. They must know how to direct, motivate, and manage their project team, contractors, suppliers, and many other individuals with whom they interact. For some this is a natural ability; for others it requires extensive training in human psychology. There are numerous sources to help your PMs improve their ability to work with people. A difficult skill to teach is leadership. Some individuals exhibit natural leadership skills—others can learn techniques to improve their leadership ability.

    Perhaps the most difficult skill to learn is that of delegation. Many design professionals and facility managers tend to be poor delegators, are ego-driven, and often lack trust in their subordinate's skills. As a result, these individuals feel the need to be in-

```
Project Management Concepts/Systems
Quality Assurance/Total Quality Management
Computers (Hardware and Software)
    • Scheduling
    • Estimating
    • Budgeting
    • Data Base Systems
    • Project Status Reports
    • Presentation Software
    • Proprietary Systems
    • CADD
    • BIM
    • Internet/Research
Contracts/Risk Management
Interpersonal/Communication Skills
    • Writing
    • Public Speaking/Presentations
    • Delegation/Motivation
    • People Skills/Managing People
    • Negotiation
    • Working with Others
Construction Inspection
Time Management
Financial Management
Project Budgeting
Scope Management
```

Figure 3.2 Project manager training topics

volved in all aspects of the project at all times. Not only does this overburden them, it hinders the performance of project team members. Learning how to delegate is a painstaking process that must be reinforced by the example of top management and by providing the tools and systems to permit adequate supervision of subordinates.

3. **Technical management skills:** To be effective, a PM must have a complete understanding of technical management skills. This covers a broad range of project activities. Project budgeting, scope

determination, staff planning, and quality assurance reviews are only a few of these tasks.

Developing Your Training Program

Most senior firm managers are inexperienced as to the how and why of an ongoing training program. This continuing education of staff and management requires a commitment, a plan, and a budget. Unfortunately, most firms leave this process up to each individual, clearly subjecting the firm's future to chance. Farsighted firm managers offer opportunities for staff and management to learn or improve their skills. Methods vary, ranging from in-house seminars to paid tuition at local colleges. Training not only improves skills, but serves as a morale booster and a fringe benefit while protecting a firm's future. Employee turnover is often the rationalization for not providing formal training. The philosophy seems to be: *Why train someone else's staff at our expense?*

One study found that 75 percent of high technology companies experienced a significant increase in employee productivity after developing and funding comprehensive formal training programs. Experience indicates similar results for the construction industry. Clearly, the cost of training is more than matched by productivity increases and is favorable toward training. In general, training goals fall into three categories:

- Teaching employees/managers how to perform a new or unfamiliar task within their current job
- Helping employees/managers improve their performance on their present job
- Preparing employees/managers to handle new jobs

Staff/Management Training

There are two major areas of training for an organization: staff and managerial. Staff training usually consists of enhancing specific skills such as computer-aided design and drafting (CADD) and product/service knowledge. Training in these areas is usually very direct, observable, and objective, and can be broken into a number of discrete parts or elements. Managerial training, however, usually focuses on communication skills, supervisory skills, and human relations. Although these types of skills are more subjective and harder to quantify and measure, they may have a greater impact on your organization. A firm may have the finest, most talented engineers, architects, or facility managers in the country, but without proper supervision, direction, and motivation, this talent may be unproductive.

Determining If Training Is Needed

Often, training is conducted for fairly limited reasons. These include teaching new skills to employees, retraining employees in skill areas they may have lost or not used in many years, or keeping employees abreast of changes in technology and design. Your first step in the training process is to determine the need for training. A thorough needs analysis should be performed on the organization by employee and position. In assessing the organization's needs, it is necessary to look at the firm as a whole. What are its strengths and weaknesses, and how does it compare to its competition?

In determining the firm's overall needs, it is necessary to look at both short- and long-term needs. For example, in a design firm, if a senior partner who has handled most of the firm's marketing will be retiring in three years, there is no time to lose in training someone to take his or her place. It is important at this level of assessment to consider the short- and long-term goals of the firm. These will have an impact on what the training needs are, or will be.

Another important aspect of this organizational assessment is the *climate* of the firm. The firm's attitude and motivational level will have a great impact on the success of any training programs that are instituted. One method of assessing the general training needs and attitude of staff is to conduct interviews and questionnaires completed by senior management, PMs, and so on.

The second level of assessment, along position lines, will help determine more specifically what and where training is needed. A thorough analysis must be done on each position in the firm to determine not only the duties and responsibilities of the position, but also the needed skills a person must have to successfully do the job.

This type of assessment involves a formal, systematic study of a position that covers a number of items. These include: what individuals in each position do in relation to information or other people; what procedures and techniques they use; the equipment and tools they need; the products or services that result from their effort; and the skills, traits, and attributes required of the person in the position.

Finally, an analysis needs to be conducted on each employee and PM to determine what skills each person has or lacks. This will help determine what training they may need to better perform their job, what position they can move into next, and what job they could grow into in the future. Assessment of employee skills may involve reviewing performance evaluations, reviewing their work, completing questionnaires, and conducting skills or ability tests.

When looked at as a whole, the identification of training needs is an involved and complex procedure. It will, however, allow an organization to assess its strengths and weaknesses, focus attention where needed, and grow in the direction desired. Few firms have a well-established training program. Most simply take advantage of isolated seminars and often, only senior management attends these programs. Training is frequently considered the responsibility of the individual, who is expected to plan, schedule, and finance his or her own program. As a result, most firms are not adequately prepared to respond to the need for new services or to meet changing market conditions.

A staff training program requires a long-term commitment and recognition that the payback may not be immediate. Regular training will result in a more productive and profitable organization. How should a training program be developed?

1. **Senior management must make a commitment to a continuing program:** A program that is conducted on an irregular basis will never achieve its goals.
2. **Establish an educational planning group/staff development task force:** The task force should be composed of three or four individuals representing all staff levels and chaired by a senior manager. This group should be charged with developing and managing the training program, researching training options and techniques, and preparing specific programs. They should meet regularly (at least once a month) and should operate on a priority basis. The Staff Development Officer should be a member of the task force (there will be more on this position later in this chapter).
3. **Develop a training plan and schedule:** This should include choosing various types of training, establishing training priorities and goals, and outlining who is eligible for each program. In addition, a schedule should be established to guide the training process.
4. **Establish a training budget:** This should be included as part of the annual budgeting process conducted by the organization. These

are funds that should be spent and not viewed as an area to cut if the firm experiences temporary declines in business.
5. **Inform your staff:** Staff members should always be informed of the various training options available and what items the organization will pay for.
6. **Require those attending outside educational programs to disseminate their information:** Sharing newly acquired concepts with other staff members can be beneficial. This could be handled at lunch meetings where short presentations are made, or in a summary report on the program.
7. **Vary the types of training programs used:** Many options are available, including the following:
    - **In-house lectures and seminars:** These programs may last from one hour to one day and they may be conducted by outside management consultants and specialty consultants, building product manufacturers, college professors, or experienced, knowledgeable staff members.
    - **College courses:** Tuition may be paid in part or in full for certain staff members to expand present capabilities or develop new ones. Distance learning courses should also be considered.
    - **Outside seminars:** The National Society of Professional Engineers (NSPE), the American Council of Engineering Companies (ACEC), the International Facilities Management Association (IFMA), the American Institute of Architects (AIA), and many other organizations sponsor numerous part-day, full-day, or multi-day courses and seminars in major cities. Many universities (particularly the University of Wisconsin, the University of California Berkeley-Extension, Michigan State, Penn State, and Harvard University) regularly offer short seminar courses of value to designers and facilities managers. Although the cost of attending many of these programs is high, they give staff and PMs the opportunity to exchange information and ideas with individuals from firms throughout the country and the world.
    - **Professional conference/conventions:** These may include conventions and conferences organized by professional design groups (e.g., AIA, ACEC, NSPE), suppliers, product manufacturers, industry groups (e.g., American Hospital Association), and others.

- **Webinars and DVDs**: Although many of these sessions and materials are too short, poorly produced, and expensive for what they offer, some may have lasting value as reference and refresher tools.
- **Resource and reference materials**: An important part of a good training program is a library of reference books, magazines, data bases, and so on. This material must be organized into a usable collection that is regularly maintained and updated.

8. **Review the performance of your training program:** At least once a year, the entire program should be reviewed for its effectiveness, cost, and impact on morale and productivity. The budget in particular must be evaluated for its short- and long-term cost-effectiveness.

Staff Development Officer

Some organizations hire a Staff Development Officer to manage the personnel operations of the firm and administer the training/education program. Responsibilities of this individual include the following:

- Assure compliance with all legal issues regarding personnel including state licensing requirements for continuing education
- Assist the chief operating officer in the development of staff benefit/reward programs
- Maintain personnel records
- Assist in the hiring/departing (voluntary/involuntary) of staff
- Administer records related to staff training/education—including Continuing Education Units (CEUs) and state licensing requirements for continuing education
- Research training/education needs and opportunities
- Contract for training/education
- Prepare a Training/Education Opportunities Bulletin to be distributed to all staff on a regular basis
- Serve as a clearinghouse for all solicited and unsolicited material on training and educational opportunities
- Help each staff member to determine their individual training/educational needs
- Assist in the development of an individualized training program for each staff member

- Serve as a member of the staff development task force
- Develop and administer the review and approval process for educational offerings
- Prepare an annual staff development budget and manage expenditures
- Develop and maintain a library of personnel, training, and educational resource materials
- Administer the tuition reimbursement program
- Help to determine incentives for each staff member to pursue training and education
- Assist in the development of career paths for staff
- Help in the creation and implementation of a mentoring program

What Generates Effective Learning?

Probably the most critical factor in determining the success of training is the motivation and attitude of the people being trained. The trainees should want to be trained and should believe that the training will have positive results. There are a number of steps that can be taken to help foster these feelings, before and during the training process. Most importantly, the goals and desired outcomes of the training must be conveyed to the trainees. During training sessions, there should be rewards for learning the material. Reinforcement must be provided for making use of what was learned and for proper learning or training behavior. Trainees will learn and remember material that they consider to be meaningful and important.

Every training program should begin with an overview of the material to be covered and an explanation of how it relates back to job problems or performance. The material should be broken into logical pieces, and these should be put into a rational, progressive sequence. The terms used and the technology discussed should be familiar to the trainees. New terms and technology should be presented in a manner that the trainees can relate to. Visual aids should be used whenever beneficial.

Cross Training

Design, construction, and facilities organizations must provide cross training for all staff members. There are three primary benefits to cross training.

- Staff develops the necessary skills in the event of turnover, vacations, illness, and so on. It is vital to maintain continuity of service. Cross-trained staff can more easily maintain your standards.
- Expanding workloads will require additional trained staff and PMs to service new and existing projects and facilities.
- Cross training allows everyone to better perform their job. Understanding why information must be in a particular form, how work is to be completed, and other team member's information requirements all help to improve everyone's performance.

Training Practices and Methods

There are a number of different methods by which training materials can be presented. The method of presentation will impact the effectiveness of the program. No single method can be used for all types of material. A training program should be designed for maximum efficiency within the constraints of time, cost, location, and equipment availability.

1. **Lecture method:** Probably the most familiar and widely used instructional method is the lecture. It is usually done live, but it may be presented on video, on audio, or on a computer utilizing a distance-learning format. From a training/learning standpoint, the lecture is one of the weakest methods. It usually involves no interaction, practice, study, or testing of the material presented.
2. **Classroom training:** Traditional classroom training is basically a series of lectures. Classroom training allows for modifications and enhancements to the lecture method by providing workbooks, small group discussions and practice, multiple sessions with homework, and regular testing. This approach is certainly more effective than a simple lecture and is appropriate for more complicated training. It can be adopted for almost any type of staff training. This could include specific technical skills such as CADD, and can be enhanced with a lab setup, using CADD terminals for hands-on practice.
3. **Programmed learning:** This is probably the best method, at least for skills training. This method breaks the training program down into many smaller parts that are put into a logical sequence. At the end of each section or module, the participants are tested and given immediate feedback as to their understanding of the material. Training courses of this type are much more difficult and

costly to develop and they usually involve programmed texts or workbooks. There are many advantages to this type of training approach. It is designed to be individually paced, with each participant moving at his or her own speed. Frequent testing can determine if the material is being understood.
4. **Group discussion/case studies:** This method is very familiar to most professionals and can be used as a separate approach in and of itself, or in conjunction with lectures. In this method, small groups discuss issues or problems, work out new ideas, solutions, proposals, and so on. It is most effective for teaching problem-solving and decision-making skills, presenting complicated or difficult material, or changing opinions and attitudes. As such, it is probably most useful for management, rather than staff-level training. It is particularly useful in human relations, communications, and supervisor training programs. Depending on the nature of the material covered, it may include intense confrontation and discussion or argument, role playing, case studies, management games, simulation, and exercises.
5. **Summary:** This material is only an overview of training program approaches. All of these methods can be used either in-house or at outside locations, but are typically conducted off the job or off-site. These approaches take the trainees away from their regular jobs—and in many cases, programs away from the work site will result in a more productive learning atmosphere.

On-the-Job Training

Another type of training method is on-the-job training (OJT). This is often used in training for specific skills. Trainees who learn while they are actually on the job are being productive. OJT is usually combined with classroom training or other off-the-job techniques as well. Internships are an example of this approach. Firms often combine OJT with other methods for better results.

Mentoring Programs

Mentoring programs match an experienced individual (the mentor) with a less experienced individual (the novice). The mentor is to guide in the development of the novice's skills and career path. A more detailed discussion of mentoring follows later in this chapter.

Program Evaluation

The purpose of a training program is to increase employee and organizational performance and productivity. The evaluation of any training should therefore focus on measuring these factors.

Managing the Training Program

A decision must be made whether to develop and provide training in-house or purchase programs from outside vendors. There are advantages and disadvantages to each approach. A primary consideration in deciding to train using in-house resources is whether or not you have the expertise and the facilities to do so. Another major consideration is your budget. There is a cost to providing training. As with any other project, costs need to be determined and a budget established. The overall budget must cover a number of items. These include training materials and supplies, facility use/rental, instructors' salaries, price/cost per trainee, plus the loss of productivity while the trainees are off the job. As a general guide, a design or construction organization should consider spending at least 5 percent of its annual total revenues on the various forms of training.

Consideration must be given to a program's timing as well to determine what month, week, or days of the week are most convenient. The location and facility must be chosen. Participants, supervisors, and managers must be notified of all details. Some of the steps mentioned can be eliminated if a decision is made to go with an outside training provider. However, careful thought should be put into evaluating and choosing outside providers.

Sources/Providers

There are literally many thousands of outside providers of training programs for design and facilities organizations. Most, if not all, professional organizations either conduct or sponsor professional training. Other providers of training services are management consulting firms, industrial psychologists, and colleges, universities, and other schools. Numerous books have also been published on the subject of training. Check Amazon.com, bookstores, or your local and university libraries for titles on this subject.

Training of young staff should begin immediately upon graduation. The AIA has developed an excellent method called the Intern Development Program (IDP). This program provides a structured framework that exposes young, unlicensed architects to the specific areas of practice required to pass state licensing exams and eventually contribute to their employers' practices. The IDP also provides for special advisors and offers a series of study guides covering all areas of architectural practice.

MENTORING PROJECT MANAGERS

Mentoring Programs

One of the most effective devices to train and develop aspiring PMs (or any staff for that matter) is through a mentoring program. This device will likely achieve the fastest development of key staff and will enhance the performance of all who participate. Younger staff gain the benefit of the experience and wisdom of older hands and the more senior individuals are exposed to new ideas and techniques.

While mentoring seems to be a *win/win* situation for all involved, it is not always embraced in design and owner/client organizations. Depending on how extensively the program is developed, there can be sizable costs for staff time, administration, training fees, lost billable time, and many other items. Older staff may resist mentoring newcomers for fear of job security. Senior management may ask, "Why pay the higher salary of the older staff when the younger individuals can do almost the same (or the same) job?" Personality conflict can also be an issue if the assigned mentor is unable to achieve a good working relationship with the individual to be mentored.

In some situations, firms are strongly in favor of establishing mentoring programs; however, sometimes the younger staff members fail to appreciate the benefits of the program or the opportunity being presented to them. A number of years ago, as the chair of the Chicago AIA chapter Practice Management Committee, I helped to establish a mentoring program. A chief goal of the program was to match up architectural practitioners in Chicago with fourth, fifth, and sixth year architecture students at the local architectural schools. Practitioners were grouped by their specialty such as design, technical areas, computer technology, and management, and students could select from their area of interest. Many practicing architects were enthusiastic about the program and readily signed up. Despite heavy promotion at the three local architecture schools, very few students applied for mentors. Sadly, this was a lost opportunity to learn a great deal about their chosen profession from those with real-world experience.

Types of Mentoring Programs

Mentoring programs exist in many forms; however, they can be broadly classified into two major groupings.

1. **Formal mentoring programs:** A formal mentoring program requires a significant commitment of resources. Extensive record keeping is required, training budgets must be expanded, administrative systems for the program need to be developed, and lost billable time must be anticipated and monitored. There are typically four steps to the development of a formal mentoring program:
    a) **Description:** Each job/position in the organization must have a written position description. Career paths must be laid out for each job grouping such as *project managers*. The criteria for advancing into a particular job/position needs to be described, and the criteria for performing a particular job/position must also be outlined. This step requires the organization to develop a process for mentoring. The firm must evaluate each position, staff advancement practices, training needs, and so on. It also provides benchmarks to measure performance and goal setting. For staff, this step provides a road map for advancement and improvement. Obviously, the achievement of goals must bring an appropriate reward such as a promotion or salary increase.
    b) **Skill assessment:** Less experienced individuals participating in the mentoring process must first be assessed as to their current skill level and performance. This provides a benchmark for them and for the firm to measure their growth. Mentors must also be assessed for their capabilities in meeting the requirements of their role.
    c) **Mentoring:** The actual mentoring process is time-consuming and continues over a long period of time. Implementation may require less experienced staff to seek an advanced degree, take other college courses, attend professional development seminars, read books, or undertake a myriad of other activities. The mentor must monitor and encourage this effort. Eventually, a mentor's role may end and a new individual may take over as mentor.

d) **Performance assessment:** The success or failure of the mentoring effort must be regularly assessed. This process must be on an individual basis and for the program as a whole. The mentoring program must be continually appraised and adjusted to reflect the needs, capabilities, and growth of your staff.
2. **Informal mentoring programs:** This is the most common type of mentoring program and is widely practiced in the construction industry. Informal programs occur in two types.
    a) **Kismet:** In this type of informal program, mentoring is left to fate as younger and older staff develop friendships and working relationships. In some cases, a more experienced individual is committed to the development of junior staff and seeks to help them advance. Generally, the firm plays little or no role in this mentoring process.
    b) **Active encouragement:** In some firms, the informal mentoring process is developed by encouraging more experienced staff to mentor junior members of the team. Activities may be planned to foster this effort, budgets may provide funds for mentoring and training programs, and a general atmosphere of concern for the development of younger staff may pervade the organization.

4

PERSONNEL PLANNING AND MANAGEMENT

PERSONNEL PLANNING

General Comments

While much of the personnel planning information that follows is directed toward design firms, many facilities management groups have similar needs. Effective project management requires tight control over personnel levels and use. The room for error is even less in small design firms as the loss or gain on a single project can result in great over- or under-staffing and thus seriously impact firm profitability. Clearly, hiring and firing staff as levels of work rise and fall seriously affects productivity and requires the constant training of new people. Maintaining staff at a planned level requires not only a consistent marketing effort, but also necessitates regular forecasting of present and future workloads.

Personnel planning is not an exact science. The key to its success is the regular discipline of preparing summary plans. A computer spreadsheet is not essential, but it can help store and manipulate data. Human judgment is the most important element of an effective plan. The input to the plan should come primarily from those actually responsible for running projects. However, one individual must be responsible for assembling the data in final form. The plan should be prepared at least weekly and reviewed and modified as needed. If the firm primarily handles small projects, it may need to prepare a plan more frequently. It is not sufficient to mentally perform this planning exercise since this does not permit a look far enough ahead to systematically and objectively examine workload.

Figure 4.1 presents a suggested personnel planning format based upon a design firm of seven technical people. Note that this firm may actually have a total staff greater than seven, but for planning purposes only technical staff or full-time equivalents are used. Full-time equivalents are part-timers converted to fractions of regular staff based on a 40-hour work week. Thus a 20-hour-a-week intern equals one-half of a full-time person. (The same calculation could be performed based upon annual hours worked using 2080 hours for an average work year without

Firm:	ABC Designs
Date:	12/15
Current Personnel Level:	7
Chargeable Ratio	0.85
Chargeable Hours Per Man Month:	0.85 X 170 = 145 (thus 1015 available without overtime)

Project Number	Staff Hours to Start	1st Month	2nd Month	3rd Month	4th Month	Staff Hours Left
In-House Projects						
0620	1,400	200	200	250	250	500
0615	450	100	150	200		
0712	1,100	200	200	200	200	300
0897	200	50	100	50		
0856	2,000	300	400	400	500	400
0843	1,000	100	300	400	100	100
Misc.	100			50		50
Subtotal	6,250	950	1,350	1,550	1,050	1,350
Probable Projects						
BD237	1,500				150	1,350
BD314	450					450
BD256	1,200					1,200
BD295	300					300
Subtotal	3,450				600	2,850
X 75%	2,588				450	2,138
Grand Total	**8,838**	**950**	**1,350**	**1,550**	**1,500**	**3,488**
Equivalent Employees		6.6	9.3	10.7	10.3	
Personnel Deficiency Surplus		4.0	2.3	3.7	3.3	

Figure 4.1 Personnel planning

overtime.) If administrative staff or others perform some technical work such as specification preparation (and this is charged to projects), it may be desirable to factor them into full-time equivalents.

The chargeable rate is the percentage of time actually available to work on projects (without overtime). Twelve percent of total time is normally lost to vacation, sick leave, holidays, and personal time off. In addition, when calculating a chargeable rate for total staff, a significant amount of time is normally used for overhead-related items, reducing the overall firm chargeable rate to the range of 60-65 percent. The data for calculating chargeable ratios must come from historical records. The actual hours available per employee in a month is calculated based on 4.25 40-hour work weeks per month, times the chargeable rate, resulting in 145 hours available per person. For a seven-person staff, this results in 1015 hours available per month without overtime.

In general, a personnel projection should be performed for the current month plus at least three additional months. Beyond a total of four months, the projection may become inaccurate except for very large projects with a predictable work flow. In Figure 4.1, the *Staff Hours to Start* column should reflect the project manager's (PMs) estimate of actual personnel time required to complete the work. It should not simply be the hours remaining in the budget, unless by chance they are the same.

The *In-House Projects* section reflects those projects currently under contract by the firm. The *Probable Projects* section includes projects that are in the business development phase, including those in the negotiation process. The hours under *Probable Projects* are based on a preliminary scope evaluation or on a best guess. A weighting factor of 75 percent is applied to adjust for unforeseen changes. This weighting factor is for illustration only and should not be used by your firm.

Each month is totaled to arrive at the total number of hours required by all projects. This total is compared with the total available of 1015 hours to arrive at a total for equivalent employees. For example, in the first month, 950/145 = 6.6 employees are required, leaving a surplus of .4. As a result, the firm is properly staffed to complete its work without overtime. In the second month, the firm begins to be substantially short of staff (2.3) requiring some adjustment. As part of the analysis of the required staff time, there must be an evaluation of the skills required to perform the work. This often significantly complicates the personnel planning format shown in this section. While the basic principles remain the same, a detailed spreadsheet may be required to perform this analysis.

Leveling Workload

Tied to forecasting personnel requirements is the need to level the workload for the available staff. This can be accomplished in several ways:

1. **Work overtime:** Although this is the obvious solution to short-term work increases, it can be counterproductive in the long run. Studies have shown that generally working 20 percent or more overtime for a sustained period (two weeks or more) can result in a significant drop in productivity. This will often negate any gains achieved by working more hours.
2. **Improve staff productivity:** A highly productive and motivated staff can readily deal with short-term work increases. A good working environment using appropriate computer technology can often greatly enhance productivity or allow the examination of additional options. In addition, keeping staff motivated by providing bonuses and other incentives as well as keeping them informed can be very important.
3. **Control the timing of discretionary time off:** A regular staff planning process can help predict workload bulges and slack periods, allowing appropriate scheduling of discretionary time. Vacations can be scheduled for slow periods and discretionary sick leave (e.g., for elective surgery) can be anticipated and matched to workload.
4. **Hire/fire short-term staff:** Many firms hire some staff on a project or other short-term basis. This avoids the damaging psychology of the regular hire/fire approach by allowing short-term staff to plan ahead and seek other, permanent employment while still drawing a regular salary. Some senior staff people who have left other firms prefer this approach since they can seek their desired position while still working, even if it is at a reduced salary. Small firms often gain a great deal from having this expertise available for even a short while.
5. **Farm out work to other firms:** During very busy periods, it may be possible to shift some work to other noncompeting or friendly firms. This may be particularly true for working drawings, and is often advantageous for firms with temporary increases in work or where the additional staff cannot be located or hired in time.
6. **Find alternative activities for your staff:** When a firm's workload temporarily declines and it wishes to keep the staff together,

alternatives are available. For example, many firms find marketing activities for their staff to perform. This might include the preparation of graphic materials, writing newspaper or magazine articles, and performing research. Alternatively, staff can be loaned to other firms that are busy or they may assist these busy firms by performing the work in your office.

The key to maintaining a firm's financial health is to control labor costs and this requires a regular process of personnel planning. The plan is the basis by which a firm can take prompt action to increase or decrease needed staff.

Contract Staffing

The National Society of Professional Engineers (NSPE) publication, *Engineering Times* (February 2000) in an editorial noted: "Gone are the days when engineers would march straight from the graduation podium to their first job and stay for 30 years. Many of today's engineers need more than one professional home to slake their thirst for experience. Some are even hooking up with temporary staffing agencies to increase their marketability and gain exposure to a greater variety of challenging engineering projects." While no hard statistics exist for measuring the number of engineers going the contract-staffing route, anecdotal evidence indicates the trend is increasing. Not only are engineers becoming contract staff in growing numbers, so are architects, general contractor's staff, owner/client staff, and many others in the construction industry. What exactly does contract staffing mean? For employers, it means not having employees on your payroll. For contractors it means independence or self-employment, with both the positive and negative aspects that entails. The advantages and disadvantages of contract staffing differ whether you are the contractor or the employer.

Contract Staff

Anyone who has ever been self-employed understands the risks and benefits of independence. You are responsible for your own health insurance, long-term disability insurance, taxes, and record keeping. One of the most difficult issues for many self-employed contractors to deal with is the insecurity of not having a permanent job. Of course, anyone in this industry who has been through one of our periodic construction recessions/depressions understands how permanent traditional employment is. The young engineer who has been laid off two or three times in a few years soon learns that lifetime employment is highly unlikely. Even more difficult is the role of the older individual who has a more expensive lifestyle or income requirements and expects a higher salary and fringe benefits commensurate with his or her experience. I recall receiving a telephone call from a 60-year-old architect who neither could afford nor wanted to retire and was looking for contract work because he was unable to find permanent employment.

Independent contractors must always be alert for work opportunities. After the project on which they are working is completed, they may be out of a job. If workload in their employer's firm declines they may need to find another employer. Admittedly, this may be true for anyone working in our industry. Despite the drawbacks, being an independent contractor has benefits. You can use your freelance status to hop from job to job or project to project in order to gain experience, travel to and work in new locations, and try out different opportunities. Contractors are often paid more per hour than employees—as employers do not have to pay the traditional 25-40 percent in fringe benefits and can therefore afford to pay higher hourly rates to independents. Those with special skills may be able to work for the highest bidder and set their own rates.

Self-employment allows tax deductions not normally available to employees such as home office space, supplies, equipment and auto leases, and other business expenses. Obviously, you need to confirm these deductions with your own accountant and it will vary based upon your particular circumstances. While you may have to pay for your own health and related insurance, you may be covered under your spouse's policy and you may not have been taking advantage of coverage while you were a traditional employee.

Contractors may also have more freedom during the business day and week. You may be able to set your own hours or days of the week to work. Employees rarely have that option. Retired individuals look at contract work as an opportunity to remain active, keep their skills up-to-date, and earn additional income. A few years back, I met a former Michigan Department of Transportation (DOT) engineer who had retired to Arizona and became bored with the inactivity and went back to work on a contract basis with a local county DOT. Young people often look at contract work as an opportunity to see if they like a particular aspect of their chosen career, experience a variety of roles, or to find out if they like working for a particular firm before accepting a permanent position.

Employers

Some of the benefits to contractors can also be disadvantages to employers. Contract staff can leave and go to the highest bidder. They may choose to depart at inopportune times leaving you without a key player on your project team. Of course, the same may happen with traditional staff. Your ability to supervise, direct, or discipline contractors may be limited as they are not your employees. The United States Internal Revenue Service (IRS) may also rule that if they are under your direction and control they may be considered employees subject to all appropriate withholdings. You must be careful to understand the rules related to independent contractors.

The advantages to using contractors far outweigh any negatives. If you need to train some of your less experienced staff, but cannot afford nor want to commit long-term to an experienced senior individual, then an independent contractor may be the answer. Employers with a temporary short-term workload surge can retain contractors to meet this need. If a particular skill is needed for a project, retaining a contractor with this experience may be the best solution.

Contract staff levels can be quickly reduced if workload temporarily or permanently declines. Fringe benefit costs can be cut greatly. Your firm does not need to supply health insurance, disability, or other voluntary benefits. Mandatory/statutory benefits such as unemployment insurance, workers' compensation, and the employer's portion of FICA (Social Security) are not required for contract staff. These individuals do not participate in your profit sharing, 401k, or bonus programs. However, you may need to increase or enhance your general liability policy when you have contract staff regularly on your premises.

Some owner/client firms are moving toward contract staff for their facilities managers (others are completely outsourcing the function altogether). In some cases, these individuals are former employees who were offered the choice of becoming contractors or face dismissal. Obviously, many accept contract status and perform the same job as before, often at a higher base rate, but without fringe benefits.

Summary

The use of contract staff (and outsourcing) will continue to grow throughout the construction industry. As *Engineering Times* noted: "Contract engineers help companies better manage the ebb and flow of project-related work and bring in appropriate specialists that may not be available among the core staff for a specific project. Lower recruiting costs and the advantages of having a flexible segment of employees are also reasons for engineering companies to use contingent labor."

STAFF MANAGEMENT

General Comments

The hallmark of a successful PM is his or her ability to effectively delegate assignments to other staff members. Growth in many organizations is inhibited by the inability of the firm's senior owners and managers to release the reins of control over project and firm management. As a result, crisis management prevails as the overextended individual attempts to cope with a work overload and shifting priorities.

Techniques for Proper Delegation

Many engineers and architects have been slow to learn the skills of effective delegation. Often those individuals who try to delegate do so improperly and are likely to be disappointed by the results. Other designers, either unable or unwilling to delegate, take the attitude that it takes more time to explain a task than to perform it oneself. In many successful and profitable companies, managers constantly find opportunities to train staff to handle delegated activities. Proper delegation requires a series of steps. The five steps listed here cover only some basic ideas. Each PM will need to determine those methods most effective for their own situation.

1. **Select the right task to delegate:** Frequently, managers assign to subordinates tasks they find undesirable. As a result, they conduct little follow-up to see if the assignment is being correctly performed. Generally, a task should be delegated when the manager's time is more profitably spent in another activity. When selecting a task for delegation, evaluate the skill and experience of the individual receiving the assignment. If additional training is required, be prepared to devote the time and effort or do not delegate the task. Repetitive tasks are ideal for delegation as are situations where a less experienced individual is being mentored. Once an assignment has been delegated and the task learned, the manager is free to perform other activities.

2. **Select the right individual to perform the delegated task:** Some tasks are delegated to individuals who find the assignment as uninteresting or unchallenging as the manager. As a result, they either postpone the task, perhaps to a crisis point, or simply delegate it to another (third) party. This latter situation may result in the assignment being performed with inadequate supervision. The result may not be satisfactory to the original manager.

   In other situations, the assignment may go to the most available individual without consideration of the skills required. Often, managers delegate tasks to their already overloaded support staff or administrative assistants. As a result, proper attention cannot be given to the assignment. A successful manager tries to match assignments with skills and time availability. He or she should provide the necessary training if skills or experience are lacking. When an individual is overloaded with other work, the manager should assist in developing priorities and schedules.

3. **Give the assignment correctly:** There is a tendency on the part of many busy managers to simply delegate an assignment without providing adequate instruction. The result is that the individual receiving the assignment may waste time in doing the wrong task or may need to continually ask questions. This defeats the original time-saving purpose of delegating the job. When an assignment is not given correctly or completely, the final result may be inadequate or may lack important information.

   Many managers believe that the time required to explain an assignment justifies their rationale that they might just as well do it themselves. This rationale fails to consider both the alternative use value of their time and the educational benefit often gained by the individual to whom the assignment is given. You should not provide step-by-step instructions, only expectations, general guidelines, suggested resources, and so on.

4. **Set a time limit or deadline for the task:** It is a wise manager who sets achievable deadlines in advance of the real deadline. This allows time for corrective action, if necessary, and offers the opportunity to provide additional training to the individual receiving the assignment. Failure to set a deadline implies that the assignment is either not important or may be postponed. A deadline that fails to allow opportunity for corrective action can create a crisis situation, potentially damaging to the manager's immediate effectiveness.

5. **Provide a review and control mechanism:** When delegating an assignment, it is vital to evaluate how well an individual is performing. This allows an opportunity to ask questions, conduct training, and review performance. Review sessions also provide an excellent opportunity to take or direct corrective action.

Responsibility and Authority

One of the most difficult concepts for firm and PMs to implement is the equality of responsibility and authority. In many organizations, project, marketing, and financial managers are assigned significant levels of responsibility. They are expected to meet budgets, deadlines, and targets—often without adequate authority to implement their decisions or to meet the needs of their tasks. This is particularly true of facility managers whose input is often ignored when new or redeveloped facilities are planned.

A common complaint in many small and mid-sized design firms is that middle managers fail to take responsibility and initiative for their assignments. Often, this occurs either because responsibilities are not clearly defined or because responsibilities are delegated while authority is not. Where authority is delegated, the residual right to override middle management decisions may remain with design firm senior management. This problem is particularly apparent in firms where the founding principals or partners are still active and are accustomed to making all decisions. These entrepreneurs often feel the need to be highly involved in marketing, project decisions, and client meetings. As a result, they may inadvertently or by habit discourage the taking of responsibility.

Delegating Responsibility

Effectively delegating responsibility requires several steps:

1. **Clarify and define exactly what each individual's responsibilities include:** This requires writing a position description and developing a detailed outline of specific task assignments.
2. **Train individuals in the use of tools and techniques necessary to meet their responsibilities:** Some employees fail to assume responsibilities because they lack the necessary skills. This might be in a technical area, but more often it is due to lack of experience

in communication skills, organizational techniques, or the management of people.
3. **Allow individuals the opportunity to do tasks their own way:** Unfortunately, some senior managers who have been doing a task a particular way have difficulty allowing subordinates to learn through their own errors or to develop their own way of doing things.
4. **Use restraint in supervising the work of subordinates:** Senior managers should teach, not dictate. They must also learn to encourage those with specific responsibilities to find their own solutions wherever possible. Second-guessing is a sure way to destroy a subordinate's decision-making ability. As a result, he or she will fail to take responsibility for assignments.
5. **Delegate sufficient authority to perform the assignment:** In general, to adequately perform tasks requires at least rough equality of responsibility and authority. Any significant imbalance in this equation will seriously handicap decision-making ability.

5

SOFT SKILLS FOR PROJECT MANAGERS

INTRODUCTION

General Comments

The focus on these soft skills should not be treated as an afterthought and given lower priority in the discussion of the role of project managers (PMs). These skills, particularly communication skills, are of great importance. Therefore, they are covered at an early stage of this book. No PM can be effective without mastering all of these soft skills. We begin by providing some ideas on relieving the time pressures most PMs face on a daily basis.

TIME MANAGEMENT

Many PMs complain about the lack of time in their lives. Crisis management makes their time management inefficient, and their families often pay the price for their absence due to job demands. One important solution is the reorganization of the firm's internal operations into a more efficient system. Many design firm principals attempt to manage projects day-to-day, as well as the operation of their practices. Few individuals can handle both effectively. Many owner PMs and facility managers find their time consumed by endless unproductive meetings. Short of a reorganization of the firm, there are many other ideas and approaches that can be used to improve time management, including the following:

1. **Control the number of people you interact with each day.** This can be achieved in part by holding regularly scheduled meetings—this avoids ad hoc meetings. Encourage decision making at the lowest effective level. By preventing too many decisions from being sent on to higher-level managers, bottlenecks are avoided. Use technology appropriately to share information, ask questions, etc.—however, do not inundate people with email and text messages.

2. **Try using a quiet hour in your office.** The goal of this concept is to create a period of time each day for quiet office work that involves concentration. Many designers and owner PMs now achieve this result by coming in very early in the morning, staying late at night, or coming into the office on weekends. This strain on family life and on the individual can be lessened by using the quiet hour. Firms using this concept set aside an hour each day free of meetings, interruptions, and telephone calls. Your assistant should take messages or send the caller to voice mail, except for emergency calls. Intraoffice calls are not permitted and you cannot visit with anyone else in the office. Nearly all firms that use the quiet hour eventually abandon its use simply because it is not enforced and individuals do not respect each other's time.

3. **Improved delegation can improve time management.** The hallmark of a successful manager is the ability to appropriately and effectively delegate an assignment to other staff members. Many in the construction industry have been slow to learn the skills of delegation. Frequently, those individuals who try to delegate do so improperly and are disappointed by the results. Others, either unable or unwilling to delegate, have the attitude that it takes more time to explain a task than to perform it themselves. In many successful and profitable companies, managers constantly seek opportunities to train staff to handle delegated activities.

4. **Managing meetings more effectively can improve your time management.** Designers in particular seem to hold endless meetings. This may be a result of their desire to reach decisions by consensus, or simply because the need for meetings may be inherent in the business. Whatever the reason, it is important to learn how to run an effective meeting.

Meeting Management

Specific suggestions to improve your meetings include the following:

1. **Hold regularly scheduled meetings:** Rather than deal with an endless series of ad hoc meetings, wherever possible, hold topics until your next regularly scheduled meeting. Typically, these include weekly marketing meetings, project team meetings, biweekly firm management meetings, weekly PM meetings, and weekly or biweekly client meetings.
2. **Prepare an agenda and stick with it:** Distributing an agenda well in advance of a meeting serves as a reminder of its time and purpose, encourages advanced preparation, and provides a framework for discussion. The agenda aids in preventing the meeting from going off on tangents. Agendas should have specific times listed to discuss each topic. This allows those who only need to be in attendance for a particular topic to come at the appropriate time.
3. **Apply effective time-management techniques to your meetings:** Start your meetings at the designated time even if several individuals are missing. Although a quick review may be necessary for those who are tardy, it is better to accomplish something than sit and wait for stragglers. In your minutes, list not only those present and absent, but also those who were late and by how long. (Some firms fine those who are late.) This latter technique is likely to be more effective for internal meetings rather than those with outsiders. Discourage small talk and extraneous conversation, as this disrupts the meeting and is a time waster. Don't allow any interruptions for telephone calls and staff questions, as these waste time and disrupt the flow of the meeting. Always set a time limit for your meetings and wherever possible, stick with it by conveying a sense of urgency to the time scheduled.
4. **Use minutes and notes effectively:** The minutes not only record discussions, but also what decisions were made, what actions are to be taken and by whom, set deadlines, and ensure that all attendees and interested parties have a complete and accurate transcript of the meeting. In order to ensure prompt action following your meeting, distribute what is known as *instant minutes*. These are nothing more than a photocopy of the minute-taker's notes.

Without this immediate encouragement to action, many individuals wait for the formal minutes to be e-mailed. This delay (often 1-2 weeks) may result in informal meetings to refresh memories, review decisions, and so on. Formal minutes are important and should follow your *instant minutes*. Include an executive summary at the top of the formal minutes.
5. **Manage your meetings effectively:** Always have a chairperson. Make sure everyone comes to the meeting prepared by distributing the agenda well in advance and by calling or e-mailing everyone to review what is expected of them at the meeting. If it is obvious that the attendees are not prepared, immediately set a new date and promptly adjourn. At the beginning of your meeting, review the agenda, summarize the results of past meetings, and emphasize the time frame allocated. At the conclusion of the meeting, review key decisions, actions to be taken and by whom, and due dates.
6. **Only invite those who need to be there:** Meetings are not a substitute for effective communications. Inviting everyone who may have even a passing involvement in a project is a waste of their time. There are better ways to communicate information to others.

Other Ideas

There are many other effective time management techniques. It is important to find the ones that work best for you. For example, your phone or other electronic devices can be helpful in planning each day and future meetings. Make a list of things to do each morning and cross them off as they are accomplished. This gives you a measure of your progress during the day.

Telephone Time Management

We all need the telephone. It is one of our most effective communications tools and source of information. The smartphone can be a major time waster and it can be disruptive to your workday. Business calls, e-mails, text messages, unwanted solicitations, and responses to your inquiries all interfere with the effective use of your time. Even a short call can disrupt your thought processes, causing a need to refocus on the task at hand. With some effort, you can make it less disruptive and more effective.

Some suggestions to make better use of your telephone time management include:

1. **Use voice mail systems:** The great advantage of voice mail is the ability to give complete and specific information to the listener. This can be a tremendous time-saver and can prevent misunderstandings and confusion. Unfortunately, some firms with voice mail systems only provide 30-60 seconds of recording time. Smart firms allow unrestricted time periods. The best voice mail systems are designed so that a *person* answers the telephone and connects you to voice mail if necessary. Poorly designed systems have an electronic voice answer first and offer menus or a directory of extensions (your extension should be printed on your business card). Voice mail also allows after hours access or communication when you are out of the office or can't be reached on your cell. Of course, voice mail systems provide an excellent way to screen calls or to avoid constant interruptions while not missing an important call or information.
2. **Train your assistant in proper telephone techniques:** Assistants and other support staff can cost you time when they should be saving time. Make certain they understand how to take a proper telephone message. A simple name, time, and return number message is totally inadequate. They should try to assist the calling party and save you time by asking what the call concerns, offering assistance, and inquiring as to the best time for a return call. Support staff must learn to judge which calls are important to you and which are not. Let them know when you are expecting an important call and what action to take. Keep your assistant informed of upcoming meetings and other office activities. This allows them to suggest another individual to assist the caller in the event that you are not available.
3. **Batch your return of calls:** If you wish to work uninterrupted during the day, have your assistant or voice mail system take messages for most calls. Set aside a period of time during the day to return these calls.
4. **Leave complete messages:** You should leave not only your name and telephone number, but the reason for your call, best time to reach you, and what information you are seeking (if any).

Telephone/Voice Mail Lesson

Several years ago, I attempted to telephone a friend who had recently started work at a firm in the Chicago suburbs. Calling a few minutes after 5:00 p.m., I was greeted with: "Thank you for calling ABC Architects. Our office hours are 8:30 a.m. to 5:00 p.m., Monday through Friday. Please call again." That was it; no options, no menu, no way to leave a message. I was amazed and appalled. Not long after that, I conducted a workshop where several staff members of this particular firm attended. I told this story (without mentioning the firm name) and during a break privately told them that it was their firm I was referring to. They told me that it actually cost them a major client. The client was already upset over other issues. He called one day shortly after 5:00 p.m. and was greeted with the same message. Incensed, he quickly sent off an e-mail firing the firm. This was a very significant price to pay for a poorly conceived voice mail system.

PREPARING AND EDITING WRITTEN MATERIALS

Introduction

PMs regularly prepare a wide variety of written materials. Reports, letters, proposals, change orders, e-mails, faxes, and memos are only a few of the types of written communications prepared by PMs. Most are hurriedly written, disorganized, filled with poor grammar, lack coherence, and fail to effectively communicate the writer's thoughts. Many PMs attribute their failure to write properly to a lack of time, little or no training in effective writing, and a disinterest in improving their writing and editing skills. Poor written communication can lead to devastating consequences. Project errors and omissions, disputes, poor client service, legal proceedings, and many other undesirable results can flow from poorly written materials. Most PMs would greatly benefit from attending effective writing classes and from regular writing and editing practice.

While not a panacea, the following steps will help most project managers and other individuals improve their writing and editing skills:

1. **Organize your thoughts:** Take a few seconds to organize your thoughts. What are you trying to communicate and how can you best do it? For most people, a brief outline of key items is sufficient. Don't worry about proper outlining form. Just get your

thoughts down. When finished, take a quick break, come back to your outline, and see if you have missed anything important. As you write your first draft, new ideas may come and the outline may be modified as needed.

2. **Write your first draft:** For most, this is the toughest step. Put your thoughts into words—nothing else matters. Proper sentence structure, spelling, punctuation, grammar, and wordiness are not the issue. Your first draft will be disorganized, repetitive, and your thoughts unclear. Upon review, the opening paragraph or the opening sentence will likely be completely unnecessary. Getting your thoughts on the screen (hopefully, you are not using pad and pen) are crucial. Refinement and editing come later. Unfortunately, many PMs stop here and e-mail, fax, mail, and/or otherwise distribute this very rough version of their writing.

3. **Review and edit your first draft for content:** Does your first draft say what you intended? Does it cover all of your points? Does it make sense? Put yourself in the place of the intended reader—will they be able to understand it? An excellent technique to use at this (and every stage) of the writing and editing process is to read your words out loud. The way your words sound is often how others will read them. If it is cumbersome to speak, it will be cumbersome to read. Let someone else read your work. Even if they are not expert writers, they will be far more objective in reviewing your work than you.

4. **Revise your draft for flow and clarity of ideas:** Once you have edited your first draft, put it aside (preferably overnight), then reread it. Do the ideas flow? Could you say them more clearly? Could it be reorganized for better coherence? What could be cut that's unnecessary, repetitive, or fails to support your points or arguments? You may need to move sentences or even whole paragraphs around while deleting others.

5. **Examine your writing closely for wordiness:** At this point, you are almost finished with the editing process. Your goal is to make your writing as concise as possible. Could one word substitute for a phrase, could a phrase substitute for a sentence? This is a challenging step. For most of us, we are trying to undo years of poor schooling. Remember those 500-word book reports? Most of us were taught quantity, not quality. Abraham Lincoln knew the folly of this approach—his Gettysburg Address was only about 270 words long and five minutes in length.

6. **Do a final check:** This is your fine-tuning step. Check for spelling, grammar, and punctuation. Don't simply rely on the computer's spell (or grammar) check. The word may be spelled write (right); but unfortunately, it may be the wrong word. Send out your written words now and it's unlikely you'll be embarrassed.

Take this process seriously. Many PMs don't seem to understand that readers of their words often know improper writing when they see it. A judgment on your skill as a PM is often formed by reading your written words. Poorly written documents call into question your drawings, calculations, opinions, and professional advice. To prevent this, some firms even have professional editors available to review the PM's writing and to provide written communication training. With practice, this process becomes automatic and can be done quickly. Following these steps will make you a better writer, a more respected communicator, and a more successful PM. Figure 5.1 provides a reference sheet on basic American English grammar.

Noun—A noun is a person, place, or thing. Examples: Person—girl, boy, teacher. Place—home, school. Thing—dog, jacket.

Pronoun—A pronoun is used in the place of a noun in a sentence. A pronoun may take the place of a person, place, or thing. Examples: I, you, she, he.

Verb—A verb can tell you what action someone or something is doing. A verb can also express a state of being. Examples: Action—run, jump, think, fell. State of being—am, is, are, was.

Adverb—An adverb describes a verb, adjective, or another adverb. It tells how, when, where, and to what degree. Examples: When—today. How—quickly. Where—outside. To what degree—barely.

Adjective—An adjective describes a noun or pronoun. It tells what kind, how many, and which one. Examples: What kind—happy. How many—seven. Which one—this.

Article—The words *a*, *an*, and *the* belong to a special group of adjective called articles. They can be used before a noun in a sentence. Examples: *a* dog, *an* apple, *the* boy.

Preposition—A preposition combines with a noun or pronoun to form a phrase that tells something about another word in a sentence. Examples: from, to, until.

Conjunction—A conjunction joins together single words or groups of words in a sentence. Examples: and, but, or, not.

Interjection—An interjection can express strong feelings or emotions. An interjection can be a single word or phrase. Examples: Ah!, Oh dear!, Help!, Look out!

Figure 5.1 Basic American English grammar

PUBLIC SPEAKING TECHNIQUES FOR PROJECT MANAGERS

Suggestions

"As we know, there are known knowns; there are things we know we know. We also know there are known, unknowns; that is to say we know there are some things we do not know. But, there are also unknown unknowns—the ones we don't know we don't know." This quote is from Donald Rumsfeld speaking at a Pentagon press conference in May 2002, in response to a question regarding a possible link between Iraq and terrorists. While it may be somewhat amusing to hear a politician practice obscuration, the goal of a PM is to effectively and concisely communicate. Unfortunately, many construction industry professionals fail to follow Mark Twain's advice: "It is better to keep your mouth shut and appear stupid than to open it and remove all doubt." Effective public speaking is the art of knowing not only what to say, but also of what not to say. Technique can be learned and practiced; discretion cannot.

Specific Suggestions to Improve Your Public Speaking Ability

Becoming an effective public speaker takes practice and feedback from others. Most of history's great orators practiced, practiced, practiced. The list that follows provides some ideas to make you a better public speaker.

1. **Join a Toastmasters Club or take a Dale Carnegie Course:** Both organizations provide excellent training in public speaking. Toastmasters is an international organization with chapters in nearly every sizable community. You pay a fee to belong and to attend training programs; however, the charges are modest and the process is very worthwhile. Dale Carnegie is a long-standing commercial enterprise that runs public speaking training programs in many communities on a fee basis. Both organizations are of great value to anyone seeking to improve their public speaking ability.
2. **Do your homework:** There is a great deal more to public speaking than simply showing up at a venue. Preparation is vital. Know your subject. Keep your presentation concise and focused. Longer isn't necessarily better. Prepare your presentation as far in advance as feasible. Allow yourself time to rework it if necessary.

Winston Churchill was well known for taking more than an hour to prepare each minute of one of his speeches. Rehearse as much as necessary to make your speech flow and stay within the allotted time. Consider videotaping your presentation to provide you with a method to analyze your delivery, gestures, eye contact, and so on.

3. **Know your audience:** Too many PMs make presentations to nontechnical audiences that are filled with incomprehensible jargon and technical language. Don't talk down to your audience, but balance your presentation with necessary technical information and explanations as necessary. Consider meeting with some community members in advance to better understand their concerns, issues, and level of understanding of the subject. Be polite to hecklers even if they are rude. Your behavior will often win you support. You gain nothing by stooping to the level of hecklers.

4. **Visit the presentation room in advance:** If at all possible, never walk into a room cold. Always check out the layout, lighting, acoustics, sight lines, and seating before your presentation. If necessary and possible, rearrange the room to your liking.

5. **Interact with your audience:** Even with large groups, try to connect with your audience by moving away from the podium, table, or any other device separating you from them. Interact by moving down a center aisle, going to the edge of the stage, or entering the seating area. Learn by watching the Sunday morning preachers on television. They energize and connect with their congregations through this physical involvement. Use active eye contact. In large groups, scan the audience and look toward the rear of the room. In small groups, briefly look at each attendee. Do not read your presentation or PowerPoint slides to the audience (more on this follows).

6. **Be a good listener:** If you have a question-and-answer period, listen intently to the questioner. Look directly at them and don't interrupt. Respond to the question clearly and concisely. If you are not sure what they are asking, don't hesitate to ask for clarification. Be honest. If you don't know the answer, say so. If you only know a little, then offer what you know and indicate a willingness to find out more information for the questioner. The best presenters are often able to get their audience talking about their own concerns, needs, problems, and desires. With smaller audiences,

try to get the attendees to share with each other. They will learn as much or more from each other as they will from you.
7. **Don't overuse technology:** PowerPoint presentations are ubiquitous and often deadly boring. Far too many speakers simply regurgitate what is already found on the screen. Your audience can read. They don't need you to stand up front and review the slide. As impressive as the graphics of many computerized presentations can be, your audience wants substance, knowledge, and answers. If they want fluff, they can watch network television news programs. If you want to stand out and communicate more effectively, consider going *no-tech*.

    A study at Ohio State University bears this out. Selected undergraduate classes were taught both with and without computer-generated slides. Students exposed to material in a traditional format scored better on exams than did those in classes using PowerPoint extensively. According to the lead researcher, Andrea Huff (newspaper article "Research Suggests" appeared in the Chicago Tribune, Monday, March 2, 1998, section 4, page 2, authors are Jon Van and Jon Bigness), "When a class is taught with the teacher's notes presented in a computer (format), it seems the students feel that all the authority comes from the computer. Instead of paying attention to what the teacher is saying, they just copy down what they see on the screen. They disassociate themselves from the class and become passive observers rather than actively participating in the learning experience."
8. **Provide handouts:** Audiences like handouts. They like to leave a presentation with materials for future reference. Never project anything on a screen where the audience does not have a hard copy. The projected materials may be hard to read and your audience may wish to take notes on the hard copy provided. Your name and contact information must be on the handout, making it easy for audience members to contact you in the future.

Summary

There is a great deal more to learn about public speaking than has been covered in this material. In the normal course of their work, PMs will have numerous opportunities to improve on their public speaking skills. With time and practice, the ideas discussed here will become second nature.

SUCCESSFUL NEGOTIATING SKILLS FOR PROJECT MANAGERS

Introduction

"I need to talk to my manager. I want you to be satisfied. Give me a minute, I'll be right back." Ninety minutes later, the car salesman returns. In the meantime, you have read every issue from 1974 of *Better Tires and Transmissions* in the showroom including the feature on new eight-track tape players for 1975. You've looked at the sticker on the window of each car in the showroom and even wandered out to the service area to watch repair work being done on all of those expensive late model vehicles the dealer sold. The television in the waiting area is tuned to a talk show that is airing an old program about people who went to buy a new car and were never seen again. When the salesman finally returns, he is waving a sheet of paper and carrying on about selling you the car below the price the dealership paid the manufacturer. After all, he seems to imply, they are in business to sell as many cars at a loss as they can. He shows you the *actual invoice from the manufacturer* for the car that interests you. "If you are still undecided," he says, "I'll throw in the undercoating at no charge."

If this sounds familiar, you've been at the receiving end of *Negotiating 101*. Every auto dealer provides this training for their sales staff. The better ones are so smooth you don't even know you're being played. These techniques are consistently used in an attempt to level the playing field with customers. In reality, the auto salesperson has a weak negotiating position. You have the strong position since you can choose not to buy a car from him or her. You can choose to buy a car somewhere else. You can decide what the car is worth to you. You have power. Learn how to use it. This is the essence of negotiating. Some ideas to keep in mind include the following:

1. **Do your homework:** Bill Richardson, former governor of New Mexico, former U.S. ambassador to the United Nations, and renowned negotiator, is a strong advocate of preparation. In a *New York Times Magazine* interview, Richardson commented, "I study everything I can about my adversary, talking to people who have seen him (or her) most recently. I read about the person, look at his (or her) public statements, try to find out the person's weaknesses, what he (or she) likes, sports, try to find out his (or her) family situation. When I go in, I go knowing almost everything

about the person's character and policies" (2003). While PMs are not negotiating with a Castro or Kim Jong Un, preparation is still key.

2. **Information is power:** Far too many PMs are inadequately prepared for a negotiation session. They are busy people and often fail to thoroughly prepare by assembling budget, cost, financial, schedule, or a wide variety of other important information. Avoid giving away information needlessly. You do this by answering a question with a question when circumstances require.

3. **Have staying power:** It is important to allow time to achieve an acceptable result from a negotiation. Crisis managers lack this time and often concede simply to end the negotiation. Often, they pay the price with inadequate budgets, time schedules, and so on. Take your lead from labor negotiations. These are often marathon endurance sessions where each party has made it clear they will take as long as necessary to conclude an agreement.

4. **Have a floor or walk away price:** At some point during the negotiation it may become apparent that you will not be able to achieve an acceptable agreement on price, schedule, scope, or other factors. You have to be willing to get up from the table and tell your opponent you are no longer interested in concluding an agreement. If you are not willing to do this, you will have a significantly weaker negotiating position. This is the power of *No*.

5. **Be aware of nibbles:** This is the practice of asking your opponent for more even after you have concluded an agreement. For example, your client has accepted your price and scope proposal, but happens to mention on the way out the door that he or she still needs senior management's approval. Inevitably, a request for a change, or a request for a minor tweak will occur. This is a nibble. If you are being nibbled, obtain something in return.

6. **Concede slowly:** If you are willing to concede a point or a price, do it reluctantly and make sure your opponent knows you are making a concession. Don't hesitate to remind him or her of your previous concession and ask for one in return. Bill Richardson has noted, "You cede what you consider (the) easy territory" (*New York Times Magazine*, 2003).

7. **Negotiations are not a game:** Too many negotiating *experts* treat the process as a game. They say silly things like "Assume everything is negotiable" or "Never accept the first offer." While most things are negotiable, it may be counterproductive or an inefficient use

of time to undertake the negotiation. Accepting the first offer may be reasonable if it achieves your goals.
8. **Remember Richardson's negotiations rules:** Make friends, define your goal, shrug off insults, close the deal, and always show respect.

6

DESIGN FIRM OPERATIONS

INTRODUCTION

This chapter discusses a number of issues internal to design firm operations including profit planning, pricing models, and project design fee budgeting. A detailed knowledge of these topics is vital for design firm managers and facilities managers who work with them. Neither projects nor design firms can be effectively operated without a full understanding of these subjects. Owners/facility managers also benefit greatly by knowing how design firms operate and price their work. For example, in comparing design fee proposals from competing consultants, it is of great help to understand the model used by each firm. Profit targets, overhead rates, multipliers, hourly job costing rates, and many similar items affect how a design firm arrives at the fee they are quoting to a potential client. The U.S. federal government has long recognized the importance of this and has established guidelines for allowable profit and overhead rates. Many other organizations require designers to present and/or justify their pricing model before they will sign a contract.

PROFIT PLANNING FOR DESIGN FIRMS

In a national survey of design firms conducted by Birnberg & Associates, one question addressed the issue of the preparation of annual profit plans. Out of 152 firms responding, 108 (or 71 percent) prepared such plans. Many of these firms were large, successful companies. Unfortunately, many smaller firms fail to follow suit.

The definition of a profit plan is a *management tool for formalizing the firm's financial objectives*. The benefits of preparing a plan are many, including establishing yearly goals and providing intermediate targets throughout the year. Figures 6.1 and 6.2 isolate portions of the plan for discussion. Note that consultant and non-consultant reimbursables and the markup on reimbursables are not considered on the profit plan. The profit plan is prepared in the same manner whether you manage your firm on a cash or accrual basis (not your tax basis).

Labor

Figure 6.1 isolates the portion of the profit plan concerned with labor allocation. Labor for the firm has two components: (1) project chargeable (direct expense) and (2) non-project chargeable (overhead). Each principal's time is analyzed based on historical records and future projections of workload for the breakdown of project chargeable versus non-project chargeable. An overall average of all principals is calculated and entered as in the example (25 percent project assigned and 75 percent unassigned). The importance of maintaining complete and accurate time sheets is readily seen.

A total figure for principal's draw is prepared (exclusive of bonuses, profit sharing, and so on). As shown in Figure 6.1, this amount is $160,000 for the year. Hence:

$160,000 × 25% = $40,000 direct expense (project assigned)
$160,000 × 75% = $120,000 overhead (unassigned)

This same process is repeated for the technical staff and for administrative staff. The 30 percent project assigned for the administrative staff typing specifications and report preparation. Each heading (total, project expenses, overhead) is totaled to achieve the total labor line. The following are required to complete this section:

- Total salaries and a breakdown by principals (or partners), technical staff, and administrative staff projected for the coming year
- Estimated chargeable rates for each principal and staff member projected for the coming year

Non-Labor Costs

Figure 6.2 adds all other costs (other than labor) to the plan. Non-reimbursable direct expenses are project chargeable expenses that come out

	Total	Project Expenses	Overhead
Principal's Draw	$160,000	$90,000	$120,000
25% Project Assigned			
75% Unassigned			
Technical Salaries	400,000	300,000	100,000
75% Project Assigned			
25% Unassigned			
Administrative Salaries	100,000	30,000	70,000
30% Project Assigned			
70% Unassigned			
Total Labor	$660,000	$370,000	$290,000

Note: All figures are for illustration purposes only and should not be used as targets for your firm.

Figure 6.1 Sample profit plan (labor only)

	Total	Project Expenses	Overhead	Profit
Principal's Draw	$160,000			
25% Project Assigned		$40,000		
75% Unassigned			$120,000	
Technical Salaries	400,000			
75% Project Assigned		300,000		
25% Unassigned			100,000	
Administrative Salaries	100,000			
30% Project Assigned		30,000		
70% Unassigned			70,000	
Total Labor	$660,000	370,000	290,000	
Non-Reimbursables	40,000	40,000		
Overhead	300,000		300,000	
Net Profit (Before Tax)	200,000			200,000
Net Fees (Revenues)	$1,200,000	$410,000	$590,000	$200,000

Figure 6.2 Sample profit plan (with non-labor costs)

of your fee. This would include such items as printing, entertainment, travel, some consultants, and other similar expenses. As these expenses are all project related they are listed under direct expenses. In general, you should seek to minimize these expenses.

Overhead totals are established by performing a detailed analysis of projected expenses (including all fringe benefits) for the coming year. In Figure 6.2, it is estimated that overhead for the coming year will be $300,000 and is listed under the overhead column.

Net profit (before tax) is determined by the establishment of attainable profit goals by the firm's principals. For partnerships, this figure would be the total of the partners' desired shares in addition to salary draw and any planned distribution to the staff. A reasonable profit target is 15-20 percent of total revenues (without reimbursables) or 25 percent of net revenues. Industry surveys typically show that actual design firm profit margins are around 10-12 percent in good years and much lower during construction recessions. The example shown in Figure 6.2 indicates a target net profit of 16.77 percent.

$$\$200,000/\$1,200,000 = 16.8\%$$

Based upon the profit plan, the firm has targeted $1.2 million in net revenues for the year. Net revenues are those earned based upon your efforts only, and do not include any pass-through items such as general reimbursables and consultant reimbursables. In the event that a poor economy or other factors will not permit this level of revenues, the budget should be revised. Remember, this is a work sheet and should be recalculated based upon new information and/or conditions.

Ratios/Multipliers

Net revenues (total or gross revenue less non-reimbursables and consultants) are the basis for calculation at 100 percent and each line item is converted to a percentage of revenues.

$$\$370,000 \text{ (direct labor)}/\$1,200,000 \text{ (net revenues)} = 30.8\%$$

Each item is calculated in the same manner.

A time-card ratio or multiplier is calculated by using direct labor (i.e., total raw labor without fringes) as a base of 1 shown. All other items are expressed as a factor of direct labor. For example:

Net revenues/direct labor = $1,200,000/$370,000 = 3.24
Total overhead/direct labor = $590,000/370,000 = 1.59
Non-reimbursables/direct labor = 40,000/370,000 = 0.11
Net profit/direct labor = 200,000/370,000 = .54

If a firm's 3.24 multiplier (based on direct raw labor without fringes) is too high for the local market conditions, then overhead or other factors can be adjusted to reduce the multiplier. (Once the multiplier is changed, recalculate the profit plan to review its impact on profit.) A gross multiplier would include a factor for most consultants and reimbursable expenses. For comparison purposes, industry surveys show overhead rates to be about 1.53 (1.53 percent of direct labor) and target multipliers to be about 3.00 (3 times direct labor).

A profit plan will provide the following:

- Revenue and profit goals to aim for and measure progress against
- Information for pricing new work including a multiplier, overhead rate, and billing rates
- Chargeable rates in total and by employee
- Salary budget in total that provides a target amount for raises for the year
- Overhead and marketing budgets
- Marketing goals (total volume of work required to achieve goals)

Keep in mind that at year's end it is unlikely that all the targets indicated on your profit plan will be achieved, but as experience grows, plans will become increasingly more accurate.

SCOPE DETERMINATION BY DESIGN FIRMS

There are many methods used by design firms to determine the scope of services on a project. Many work with detailed checklists. Some use approximate guidelines and others simply use the broad definition of services offered in the American Institute of Architects (AIA) and other organizations' standard form contracts. A preferred method requires detailed analysis of the activities needed to complete the project. This list of activities can be reviewed with a client and a determination made as to those for which the designer is responsible. Although architectural descriptions are shown, minor changes could easily make the formats useful for engineers or interior designers. Figure 6.3 shows a sample work sheet that can be used for determining the project scope of service.

Scope Determination	Working Drawings Phase						
	By Architect	By Architect, as Outside Services	By Owner, Coordinated by Architect	By Owner	By Architect, as Additional Service	Not to be Provided	Method of Compensation
Project Administration							
Disciplines Coordination/Document Checking							
Agency Consulting/Review/Approval							
Owner supplied Data Coordination							
Architectural Design/Documentation							
Structural Design/Documentation							
Mechanical Design/Documentation							
Electrical Design/Documentation							
Civil Design/Documentation							
Landscape Design/Documentation							
Interior Design/Documentation							
Materials Research/Specifications							
Special Bidding Documents/Scheduling							
Statement of Probable Construction Cost							
Presentations							

Figure 6.3 Scope of services planning form. Courtesy: AIA

Typically, an architect uses this type of form to inform the client of the variety of available services. Additional services such as *graphic design* could be added based upon either client request or suggestion by the designer. Disagreements over services or responsibility for activities can be minimized by notation under the appropriate heading across the top. Separate compensation methods could be indicated as shown, although this should be kept reasonably consistent to avoid confusion. Although this is an excellent tool for determining a scope of services, this system has significant drawbacks as a management and monitoring tool. Its complexity makes filling out time sheets in such detail nearly impossible. Project status reports showing this extensive detail would be extremely complex.

Dividing Contracts

Many design firms are dividing their contracts into three or more separate parts. For example, scope determination, programming, and possibly preliminary design work are handled as the first contract—often billed to the client on a time and materials basis. A second contract is then prepared once the detailed requirements of the project have been determined. A third contract is prepared for the construction administration phase. Some clients may hire outside specialty firms such as construction managers for this last activity.

SELECTING EXTERNAL (TO THE PRIME) CONSULTANTS

The early selection of external (to the prime) consultants to join the project team is an extremely important decision. This allows their input into the scope definition process and ensures that all parties are in agreement on responsibilities and financial issues. Many designers work with consultants with whom they have a long-standing relationship. Unfortunately, these individuals or firms may not have the best technical qualifications or be best suited to meet the clients' needs.

Consultant Selection Process

Every design firm recommending other consultants to clients should have a well-established process to evaluate these firms. This evaluation process should include the following steps:

1. **Establish a resource file of capable consultants:** This should include information on their specialty, if any, as well as any other pertinent data. This material should be updated regularly by your marketing staff or principals.

2. **Develop an evaluation system for consultants who have worked on past projects:** This will provide ready reference material on their performance, staff, and methods of operation.
3. **Maintain materials and an evaluation on consultants with whom you are very familiar:** You may, however, include them on your short list without an extended evaluation. Always consider other consultants. You may find them better able to meet both your needs and those of clients.
4. **Collect information on consultants with whom you have yet not worked:** This includes the following:
   a. Have consultants provide a list of references of several other prime design firms with whom they have worked.
   b. Obtain biographical data on the individuals who will be assigned to your project. Obviously, this must include information that convinces you of their managerial or technical capability.
   c. Ask for information on the consultant's financial stability. If possible, obtain a financial statement. Can the consultant afford to add staff if necessary and provide the required project staffing? Can the consultant meet the higher professional liability premiums that the additional work may require?
   d. Obtain data on the consultant's internal management structure and project delivery methods. It is a great help if his or her method of operation is similar to yours. For example, your project manager (PM) can be more effective if he or she has a clearly designated counterpart in the consultant's office.
   e. If you have a project management manual, provide the consultant with a copy as a reference guide to your operations. Require the return of this copyrighted manual in the event the consultant is not selected. (Owners/facilities managers should also request a copy of this manual.)
   f. Gather information on the consultant's specific computer software and hardware and other technical capabilities. It is important to have compatible software and systems.
   g. Require each consultant to provide you with a *certificate of insurance* indicating the existence of professional liability coverage and the amount. Failure to do so may leave you as the deep pocket in the event of a lawsuit. (Owners/

facilities managers should also request certificates from all consultants.)
   h. Obtain a summary of the potential consultant's quality management/assurance procedures. This will minimize client problems, contractor disagreements, and potential litigation.
5. **Prepare a clear and complete contract:** Never operate on a handshake or verbal agreement. This is especially true with project change orders. Without effective change order management, you may find your external consultants undertaking work beyond or not in accordance with your client's wishes. Also, without a firm agreement on fees for changes, you may find the consultant invoicing you for an amount in excess of what you will receive from your client.

Your agreement with outside consultants must be very specific on billing and payment terms and conditions. Many prime design firms are extremely slow to pay their consultants. The prime must be prepared (under severe penalty) to pay his or her consultants immediately after receiving payment from the client. This is a particularly upsetting subject to many engineers who work with architects. Some architects use money that rightfully should be paid to their engineers to finance their own poor management. This subject will be discussed again later in this chapter.

Owner/Facilities Managers Selection of Design Consultants

Increasingly, owners/facilities managers are selecting and contracting directly with consultants to the prime design firm. This is often desired by owners as it gives them greater control over all of their consultants. If owners directly retain consultants, they must be prepared to manage the project team with their own staff or hire another consultant to do it for them. Designers also find it beneficial when the owner directly contracts with consultants as it typically reduces their own professional liability insurance cost. If an at-risk construction manager is hired, they will normally retain all design consultants.

BUDGETING PROJECT DESIGN COSTS

The preparation of design budgets and the monitoring of costs are major activities for design firm PMs. Although they are often performed separately, they are basic to a firm's profitability and success in providing effective professional services. The project budgeting process has two interdependent aspects: (1) estimating compensation before negotiating with the client and (2) determining the actual detailed project budget by project phase or activity. Neither can be adequately performed without the other.

Experience indicates that too many design firms estimate compensation on a percentage of estimated construction cost basis, negotiate that fee with the client, and then attempt to fit their actual costs into the fee maximum. There are, of course, many other methods of job costing. These include time-based methods, such as straight time card and time card to a maximum. Others are task-based, such as fees based on average labor cost per sheet of drawings, while others may be historically based, such as those using lump sum or a flat fee.

The preferred method is to determine a firm's actual costs first and then communicate the proposed fee to the client in whatever form is required (percentage of construction cost, lump sum, cost plus, multiplier, or hourly rate). For example, Engineer Smith is competing against Engineer Jones for a commission. The prospective client has retained engineers in the past and has always negotiated a percentage of construction cost contract. Smith cannot present the prospect with an elaborate method of determining fair compensation if Jones simply offers a 5 percent fee to perform the work. Smith's approach should be to determine actual costs while recording areas in which he has room to adjust his fee (perhaps by altering the scope of services he will provide). Then translate this fee into an estimated percentage of construction cost by the following formula:

$$\text{Percent of estimated construction costs} = \text{design fee (including profit)} / \text{estimated construction cost}$$

As mentioned, Jones has already set his fee at 5 percent. Suppose Smith's analysis results in a fee of 6 percent. His options include the following:

- Alter the scope of services provided
- Go with the 6 percent fee
- Determine that his time is better spent seeking other projects that will provide the fee he desires
- Take the project in the hope that he can complete it within the lower fee the client is willing to pay (without damaging quality and service)
- Remind the client that a lower fee may mean lower service

The decision is his. The key is to know his costs.

Where do firms obtain the information to determine these actual costs? There are many sources including the following:

- They could go by the firm's experience and records on previous projects of similar scope. (See Figure 6.4, *Completed Project File*, for a suggested content of this historical file.)
- They could use cost per sheet (sheet counts), but only if this information is based upon documented cost per sheet for services by project type. Using the cost per sheet from an old firm or a

---

Project number
Project name
Project address
Related project references
Referenced contracts and addresses
Owners name and address
Gross square footage
Construction costs (budgeted, contracted, final)
Cost/gross square footage (exclusive of site work)
Cost analysis (if available)
Program summarization (major areas)
Design Fee (budgeted final)
Project starting date
Construction starting date
Final completion date
Final project summary report
Consultants (including addresses, telephone numbers, e-mail addresses, etc.)
Project manager's statement (one page should cover the significance of the project, personnel involved, contractor selection process, scope of services, awards, information of public interest, potential of owner recommendation, etc.)
Project designer's statement (brief statement of the program, design constraints, objectives, materials, etc.)
Copy of the final Certificate of Payment including a list of contractors
List of subcontractors and material suppliers
List should include product manufacturers for publication and field rep's evaluation of each subcontractor and supplier (delivery time, quality of workmanship, recommendations for use on other projects, etc.)
Site Plan (8 1/2 × 11) showing building configuration
Project photos, video, etc.

Figure 6.4 Information for a completed project file (electronic and/or paper)

neighbor's is not adequate. Cost must relate to the firm's level of productivity, projects, and so on.
- They could go by the manager's knowledge of similar projects. This is often the single best source of data, if combined with recorded historical information. In many firms, the PM's private (paper and electronic) files are often more complete than the firm's central files. With staff turnover, however, much of this information is often lost.
- They could use cost-based compensation guidelines. This system, developed by the American Institute of Architects (AIA), provides an extensive breakdown of all possible services provided by a design firm. It is an excellent method to isolate the various activities you must perform and to determine a cost for each activity. Many firms have successfully used this system as an educational tool to show their clients the multitude of tasks that are required to meet the clients' needs. Engineers and others can easily use the forms by simply changing the descriptions.
- They could go by a percentage of construction costs, only if it is based on the firm's actual recorded experience with that building type, with similar clients in similar locations. This method may expose the firm to excessive loss if costs run higher than the fee or if the construction cost is reduced (either by redesign or low bids). At the same time you should not try to make a windfall profit on a higher than anticipated construction cost. Your job is to help reduce costs for clients/facilities managers, not increase them to enhance your fee.

Several other considerations are important in developing project budgets. Most designers neglect to forward price their services. With long-term projects, the cost of labor and overhead will increase over the life of a project. Failure to anticipate these increases can easily turn a profit into a loss. It is important to determine the level of service the firm can provide based upon a client's needs and ability to pay. Clients with a greater ability to pay can receive more elaborate levels of service. A firm has to maintain quality and service, but must know when the fee will not support even minimum standards and be prepared not to seek the commission. To do otherwise is a disservice to the design firm and the client.

Project Cost Plan

A surprising number of design firms do not have a formal method of developing a project budget. Often, for small projects, a detailed breakdown may not be necessary but the basic cost areas should be considered. All larger projects (even open-ended, time card, or multiplier projects where an obligation to the client to control costs still exists) should be controlled by a formal budget. The more successful firms often incorporate techniques to provide cross-checks using several budgeting methods.

Figure 6.5 shows a project cost plan form incorporating a percent of construction cost cross-check (estimated construction cost). Other major sections include in-house services, direct (non-reimbursable) costs that come out of fees, profit, and reimbursable costs. Using detailed work sheets, client input, previous experience, and other methods, a general estimate of construction costs by major area is developed. This information is then used to determine an estimated fee based upon the firm's historical percentage of construction cost database. The estimate is used only as a cross-check with other methods of estimating compensation.

Services provided within the firm are determined by a careful analysis of the scope of the project services. (Figures shown are for illustration only). Hours to complete each task are estimated and multiplied by the average raw labor rate (not the direct personnel expense [DPE] factor [labor plus fringes]) of all individuals who will be working on that phase. Each of these line items is totaled to obtain the total labor amount shown. The firm's overhead is determined by multiplying labor by an overhead factor. Hence, the overhead factor line is calculated by the following:

$$\$5{,}224 \times 150\% \ (1.5) = \$7{,}836$$

Next, add project-related non-reimbursable direct costs that come out of the fee. The consultant cost budget should be determined by written agreement with each consultant. Other non-reimbursable direct costs should be determined by a careful analysis of expected costs. All costs in this section should be kept to a minimum as those costs reduce potential profit for the firm. Too often, profit is considered the amount left over when the project is complete. Profit should be planned. The first step is to determine costs by adding total labor, overhead, and non-reimbursable direct costs:

Total labor	$5,224
Overhead factor	$7,836
Direct costs	$2,600
Total	$15,660

	Project Cost Plan		
			Date: November 30, 2015
			Project No.: 02715
Scope of Project:			**Project Name:**
Provide complete design services			City Hall Alterations
for City Hall alterations			Any City, Texas

Estimate Construction Cost

1.	Site Work	$ 50,000	
2.	General	$160,000	
3.	HVAC	$ 80,000	
4.	Plumbing	$ 20,000	
5.	Electrical	$ 20,000	
6.	Kitchen	$ 10,000	
7.	Elevator	$ 40,000	
8.	Other	$ 20,000	
	Total		$400,000

Services Provided within the Firm (Sample Breakdown Only)

1.	Predesign	20 hours @ $20/hour = 400	
2.	Site Analysis	4 hours @ $16/hour = 64	
3.	Schematic Design	20 hours @ $30/hour = 600	
4.	Design Development	30 hours @ $24/hour = 720	
5.	Construction Documents	88 hours @ $20/hour = 1760	
6.	Bid Negotiation	10 hours @ $24/hour = 240	
7.	Construction Administration	40 hours @ $36/hour = 1440	
8.	Postconstruction	- hours @ $ - /hour = -	
9.	Supplementary Services	- hours @ $ - /hour = -	
	Total Labor	$5,224	
	Overhead Factor (150%)	$7,836	
	Total Labor & Overhead		$13,060

Non-reimbursable Direct Costs (Costs That Come Out of Fee)

1.	Consultant 1	$ 800	
2.	Consultant 2	$ 600	
3.	Consultant 3	$ 600	
4.	Consultant 4	$ ----	
5.	Other Consultants	$ 400	
	Subtotal Consultants	$ 2,400	
6.	Travel	$ ----	
7.	Reproduction, Supplies And Printing	$ 100	
8.	Models And Photographs	$ 50	
9.	Telephone And Telegraph	$ 30	
10.	Other	$ 20	
	Subtotal Non-reimbursables	$ 200	
	Total Non-reimbursable Direct Costs		$2,600

Profit

1.	Total Labor, Overhead, And Direct	80%	$15,660	
2.	Contingency		3%	470
3.	Profit		17%	2,662
	Total Fee For Services		100%	$18,792
4.	Markup On Reimbursables			$ 1,208
	Fee Quoted Client (plus reimbursables)			$20,000

Reimbursables

1.	Total (budgeted or by contract)	$ 5,000	
	Total Compensation		**$25,000**

Figure 6.5 Project cost plan

To this total add factors for contingencies (3 percent is a generally accepted standard) and profit (17-20 percent is generally targeted). Hence:

$$3\% \times \$15{,}660 = \$470 \qquad 17\% \times \$15{,}660 = \$2{,}662$$

The final items to consider are reimbursable expenses over and above the basic fee for service. These expenses are calculated on a separate schedule. A percentage markup (often ranging from 5-25 percent) is sometimes added to this amount. After inclusion of $1208 for a reimbursable markup, the total fee for services is $20,000. Unless specific justification can be provided, reimbursable markups should be considered a profit enhancement and thus are included in the fee for services (this will be discussed in more detail later in this chapter). Reimbursables comprise many of the same types of items (including consultants) shown under direct costs. Hence, a line item for a soils consultant may be found under both the direct and reimbursable sections as well (the same is true for other items), depending upon the contract with the client.

This completes the basics of project budgeting. After determining the actual costs, a fee may be presented to the client/facilities manager in whatever form he or she desires. To adjust the fee, modify the relevant section of the project cost plan and revise the subsequent figures.

Direct Personnel Expense

Unique to some architectural firms is the use of the direct personnel expense (DPE) factor. DPE determines the calculation of the hourly job cost rate for an employee on raw labor plus all statutory and discretionary fringe benefits. Statutory fringes such as state and federal unemployment insurance taxes, the portion of Social Security paid by the firm, and workers' compensation insurance are paid by virtually all design firms. The level of these, however, varies greatly by location. Discretionary fringes include major medical insurance, long-term disability insurance, dental insurance, life insurance, vacation, sick leave, holiday pay, and a great variety of other benefits. Not only the components of the DPE factor, but also the level of each vary from firm to firm, among employees of the same firm, and from period to period.

The use of DPE factors has long been incorporated in some standard AIA contract forms. Typically, DPE is multiplied by a factor that covers non-fringe-benefit overhead expenses, profit, and non-reimbursable (direct) consultant and non-consultant expenses. Most other design professionals simply multiply raw labor by a factor that includes all overhead costs including fringes, profit, and non-reimbursable (direct) expenses. Raw labor is defined as the dollar amount job-costed to a project even if it is different than the employee's actual pay rate.

In many cases, a firm using a DPE factor as a base will have what appears to be a lower multiplier. This may be of great value to a firm in a price-competitive situation trying to provide or give the illusion of a lower multiplier. (The use of a multiplier is particularly significant in time-card or time-card-to-a-maximum contracts.) This edge, however, may only be illusionary. It is very possible that a firm using a DPE base may actually have a higher multiplier than a firm using raw labor. For example, Firm A has a DPE factor of 1.4 (if raw labor is a dollar, then fringes are another 40 cents) and a multiplier for non-reimbursables, profit, and overhead of 2.5 times DPE. Firm B has a raw labor rate of 1.0 and a multiplier of 3.2 times raw labor. Firm A appears to have the lower multiplier (2.5 versus 3.2), but on closer examination, Firm A's multiplier is actually 3.5 (1.4 × 2.5). To the unaware client or facilities manager, Firm A appears to be less expensive, while the reverse is actually true.

In a competitive situation, it is understood that a wise negotiator will present his or her firm in the best possible light. Translating and presenting a cost structure to a potential client using DPE could be to the firm's advantage. Significant problems can be created, however, if the firm attempts to use DPE-based data for internal management and permanent records. With the fringe benefit mix varying from employee to employee, project to project, and period to period, the firm's management is never certain (without extensive research) of the actual base labor dollars required and the actual cost of doing the project. DPE-based records are nearly useless for building vitally important historical databases, as even averages are rendered meaningless by a changing package of fringe benefits. By removing fringes from overhead and including them in the DPE factor charged (usually this is done by calculating a firm-wide fringe factor to be added to raw labor), the actual overhead cost associated with a project can be distorted. Some firms avoid this distortion by providing each employee with his or her own specific fringe factor. Extensive research would still be required, however, to make use of the firm's database. The best solution is to keep all records on a raw labor base and provide clients with an equivalent DPE multiplier only when competitive conditions make it necessary.

CASE STUDY: FEDERAL DESIGN WORK

Background

The selection of engineers and architects for federal design work is governed largely by the Brooks Act (Public Law 92-582). A number of Federal Acquisition Regulations (FAR) exist to carry out the requirements of the Brooks Act. In summary, the FAR provide for the selection of the most highly qualified engineer or architect on the basis of competence and at a fair and reasonable price. While the specifics of the process will vary slightly from agency to agency, the methods used are reasonably consistent. The process often begins when a representative of a local facility requests funding for changes or improvement to an existing building or for a new facility. In other cases, agency personnel may determine a specific need. Whatever the source of the request, the process will require budgeting and an appropriation. Often, a two to three year lead time is necessary from request to appropriation.

Each agency has a number of contract officers who, according to established guidelines *must assure that the funds appropriated and apportioned are sufficient to cover the design services....* Design of new facilities, replacement facilities, and building additions must be accomplished with the funds appropriated for that specific project. Design for building improvements must be accomplished with specifically designated funds or with lump sums appropriated for repairs and improvements.

A Program of Requirements (POR) is prepared by either agency personnel or by a contract engineer or architect. The POR broadly defines, in architectural and engineering terms, the scope of the work. An architectural/engineering (A/E) Statement of Work is also prepared and outlines the scope of services required of the submitting engineers and/or architects. In general, its purpose is to avoid price negotiation problems, eliminate ambiguities, and assure that the design will satisfy the program needs. After the POR and the A/E Statement of Work are prepared, but before a public announcement is made of the need for engineering or architectural services, the agency must establish evaluation criteria. These criteria must be consistent with the FAR. It is weighted according to the importance of each factor and is listed in the public announcement of the work in descending order of importance.

Typically, criteria may include the following:

- Qualifications of the project team such as team experience on similar projects and the professional qualifications of the individual staff members who will be assigned to the project
- The engineer's or architect's ability and experience in the specific type of project needed by the agency
- The management ability and organization of the engineering and/or architectural firm—this may include factors such as the design firm's internal organizational methods for project delivery, the firm's systems to maintain quality, and cost and scheduling control ability and methods

The agency also prepares an independent, government generated estimate of the cost of engineering and/or architectural services. This estimate is prepared before the public announcement of the project. It can be revised during the negotiation process with engineers and architects if the scope of the work to be performed changes. This estimate is used to determine the adequacy of funding to cover the project as outlined in the statement of the work. It is also used to evaluate the fairness of the engineer's or architect's fee proposal. Under the Brooks Act, a statutory limit for basic design services of 6 percent of the estimated basic construction cost exists.

A government management plan is prepared describing the agency's project team, individual responsibilities, relationships and interfaces. The project officer indicates the major actions to be performed and those responsible for the actions. At this point, the agency is ready to make a public announcement of the planned engineer and/or architect selection. Interested and qualified engineers and/or architects typically respond by using U.S. Government agency standard forms with the submittal, if one is not presently on file with the specific agency. In addition, the agency may request supporting material such as the firm's brochure or other information.

Most agencies have a selection board to evaluate and negotiate with designers. The review board examines current data from eligible firms, including those responding to the public announcement. Typically, each board member reviews submitted material and rates each design firm. The rankings range from most qualified to least qualified. From these rankings, several firms are selected for interviews. After design firms have been selected for interviews, they are provided with copies of the Statement of Work and the POR.

At the interview, the design firm should present its project management team, consultants, management plan for the project, key personnel, and any other pertinent information. After completion of the interviews, each member of the selection board ranks the firms again. These rankings are then forwarded to the contracting officer for follow-up and negotiation. The contracting officer begins with the highest ranking firm.

The negotiation procedure begins when a letter is sent inviting the designer to submit a fee proposal that conforms to federal guidelines for allowable cost items. Designers must understand a number of cost items are typically not allowable under the guidelines:

- Interest expenses
- Bad debt expenses
- Insurance on principals' lives
- Advertising and promotion
- Contributions and donations
- Entertainment expenses

The invitation letter outlines how a proposal is to be prepared and the date and time for submission of the proposal. A sample draft copy of a contract normally is included. Large contracts may require special provisions. For example, if the proposal for design services is expected to exceed $100,000, a pre-award audit may be required prior to negotiations. If a mutually satisfying contract cannot be negotiated with the highest ranking firm, the contracting officer will obtain in writing a best and final offer. The officer will then terminate the negotiations and advise the design firm. Negotiations will start with the second ranking firm and continue with additional firms until a satisfactory conclusion is reached.

Exceptions to the process described here do occur. For example, the General Services Administration (GSA), who manages design and construction work for many federal agencies, developed a Design Excellence program. The GSA used this program for high profile projects such as an agency headquarters building, a major U.S. embassy located in a foreign country or for specific security considerations. In these special cases, the GSA invited highly recognized or specialized design firms to directly submit their qualifications without using the standard public announcement process.

MT&Y

MT&Y specialized in health care and laboratory buildings in the Midwest. The firm was founded in the 1920s and had numerous changes in ownership over the years. The owners at the time of the Argonne Laboratories project (which will be discussed below) were all veterans of large, internationally known design firms. They had extensive project management experience, but lacked firm management expertise. Although primarily an architectural firm, the staff of nearly 100 people included a few structural engineers, interior designers, and a quantity surveyor for construction cost control. In many areas, MT&Y was highly innovative, having pioneered computerized job cost reporting systems, computerized specifications systems, early stage CADD systems, and many other automated processes.

The firm had an excellent design reputation and regularly was recognized for its accomplishments. Many of the project managers were highly capable and eventually led their own design firms. The technical staff was excellent, with a deep knowledge of health care requirements.

The Achilles heel of MT&Y lay in the almost total lack of financial knowledge on the part of the firm's president and biggest ego. Like many architects, he often made decisions on an emotional basis, rather than a business basis. The other three partners/principals in the firm typically acquiesced to the president. Even where they did disagree, they lacked adequate financial and business knowledge on which to base their dissent.

Despite excellent project management, the firm was constantly in financial difficulty. Expensive office space was maintained and lavishly furnished. Large sums were spent on nonproductive marketing. The firm's overhead rate regularly exceeded 150 percent of raw (direct) labor. Opportunities to bill for additional services were often lost based on the decision not to charge for this work by the firm's president. Outside consulting engineers and other consultants were paid slowly, if at all, and they continually threatened to stop work.

There were many other sources of MT&Y's difficulty. A significant problem lay in the pricing of the firm's work. Many projects were contracted and billed on a time and materials basis to a fee maximum. In theory, if the project were completed within the fee maximum and the scope of services, a reasonable profit should result. However, if the firm undertook extensive work beyond the contractual scope and failed to obtain additional or sufficient compensation for this work, a significant erosion of profit would occur.

Even if all work was billed for, failure to properly price the work would result in little or no profit or possibly a loss. For example, for many years MT&Y's PMs were told to use a 3.1 multiplier on raw (direct) project labor. As a result, for $1 paid to the staff for producing the project work, the firm would bill out $3.10. Unfortunately, the firm's break-even point was $3.30. Only the occasional lump sum project completed under the agreed amount and periodic negotiated settlement of outstanding payables to outside consultants kept the firm solvent. MT&Y's president never grasped this simple concept. When the suggestion was made to revise the firm's target multiplier, he repeatedly refused on the basis that clients would not be willing to pay the increased amount. A reduction in the firm's cost structure was never seriously considered.

Argonne National Laboratory

Argonne National Laboratory is one of the U.S. Department of Energy's largest research centers. It is also the nation's first national laboratory, chartered in 1946. Argonne is a direct descendant of the University of Chicago's Metallurgical Laboratory, part of the World War II Manhattan Project to build the first atomic bomb. It was at the Met Lab, where on December 2, 1942, Enrico Fermi and his team of 50 colleagues created the world's first controlled nuclear chain reaction in a squash court at the University of Chicago. After the war, Argonne was given the mission of developing nuclear reactors for peaceful purposes. Over the years, Argonne's mission has expanded into many other areas of science, engineering, and technology. Today, the lab employs over 3,400 people, including 1,400 scientists and engineers. It has an annual operating budget of more than $760 million and occupies 1,700 wooded acres in a forest preserve about 25 miles southwest of Chicago's Loop. The site also houses the U.S. Department of Energy's Chicago Operations Office.

Argonne's research falls into five major categories:

- Basic science including experimental and theoretical work in material science, physics, chemistry, biology, mathematics, and computer science
- Specialized scientific facilities that would be too expensive for a single company or university to build and operate
- Energy resource programs to help insure a reliable supply of efficient and clean energy
- Environmental management programs
- Protecting national security

The Project

MT&Y was ranked the most qualified firm to design a specialized virus research laboratory at the Argonne complex. The firm's health care experience provided a strong foundation to effectively and profitably complete the project. Early in the process of seeking the work, MT&Y's most experienced PM was assigned to handle the job, if the firm was successful in obtaining the project. The PM selected a very practical designer and a capable project architect to assist in meeting the client's needs.

Despite the firm's apparent strengths, MT&Y faced a number of challenges. They had no experience in federally funded projects and knew little about working with a federal agency. The process was totally alien to any relationship they had previously established. In addition, they knew little of what was expected of the firm other than technically completing the work within a budget, scope, and time frame. The PM expected the process to be very similar to private sector work. No one in the firm developed any substantial expertise in working with a federal agency. Federal pricing limitations, unallowable overhead issues, budget constraints, reporting requirements, design change order processes, etc. were all beyond the experience of anyone in the firm.

During fee negotiations, the president of MT&Y overrode the PM's objections and concerns about the proposed *bare bones* design fee. An agreement was reached with the agency, but subject to audit as the contract called for a fee in excess of $550,000 (inflation adjusted number). The estimated construction cost for the project was over $10 million (inflation adjusted number). MT&Y's normal overhead rate ran about 152 percent of raw (direct) labor. Most years, the firm showed little or no profit. Its pricing model had not been updated for a number of years. The federal auditors disallowed a substantial portion of the firm's overhead, resulting in an allowable overhead rate of 103 percent. At this point, the president of the firm was faced with a decision. Should he sign the contract or take a pass on the opportunity? At the time, other work in the firm was showing signs of slowing. Despite serious objections from the PM, the president signed the contract and MT&Y moved ahead with the project. The end result was a disaster.

The design fee was clearly inadequate for the required work and the fee maximum was reached long before the work was completed. Initially, some of the best designers and technicians were assigned to the effort penalizing other projects and multiplying the financial impact. As the effort on the design and working drawings progressed, it became clear to the president and PM that the project was in deep trouble. Despite attempts, the fee could not be increased except for a small amount where the client requested scope changes. Finally yielding to reality, the MT&Y president made the decision to reassign key staff to other projects with a greater potential to achieve a profit. Service to the client deteriorated, deadlines were missed, work quality declined and losses grew. The project eventually lost over $250,000; the client's review (available to other federal agencies seeking designers) was extremely negative; and the firm was nearly driven into bankruptcy. (Note: The president never publicly accepted blame for his poor decision.)

OTHER ISSUES

Revealing Salary Information to Project Managers

Some design firm principals are reluctant to provide salary data to PMs. This is done in the mistaken belief that the confidentiality of pay rates is being protected. Unfortunately, this places an unfair burden on PMs. They are expected to manage costs on projects without knowing what these costs are. You cannot manage a project by using hours alone. It is extremely likely that meeting the hour budget will result in a dollar amount at great variance with the budget. This occurs because the mix of individuals actually working on a project and their pay rates are often at variance with those the PM had in mind when a fee was negotiated. Managing on hours fails to provide the PM with any frame of reference for controlling project design costs.

Some design firms have pay rates that differ from the employees' actual job cost rate. Standardized employee classifications may exist, such as Technical I or Technical II, where all workers in a class are job costed at the same rate. In other firms, job cost rates are rounded to the nearest dollar or $10 to promote standardization or to ease budgeting and management. Both result in a variance from pay rates. In these cases, the PM does not need to know pay rates—only job cost rates. However, senior firm managers must recognize that this variance will make it more difficult for PMs to effectively control costs.

Value Pricing

Value pricing and value marketing are certainly not new concepts. The architectural profession in particular has been an advocate of value pricing as fees have declined and competition from nontraditional providers of design and construction services has grown. Obviously, everyone wants to be valued for their knowledge and expertise. The crucial issue is perception of value. Architects and engineers naturally value their skills and knowledge highly. And, just as obviously, the marketplace is discounting their value or purchasing from competitors such as design builders, project management consultants, program managers, and construction managers.

Clearly, the marketplace is forcing design firms to price on the services needed by clients/owners and placing little credibility on value pricing. PMs are a crucial link in this process by serving clients successfully, participating in the marketing process, and by being alert to new or additional services needed by their current clients.

Reimbursable Markups

Some design firms mark up reimbursable expenses. The typical markup is 10 percent, with a range from 5-25 percent. Frequently, firms attempt to justify this markup by claiming they are incurring additional administrative, clerical, and coordination time (direct labor time by technical staff and PMs) and expense in processing these items. This claim is often difficult to substantiate, since these costs are normally charged to overhead. If they are being charged to clients as a markup, then the overhead rate charged should be reduced accordingly. Failure to do so will result in double charging clients for the same item. If the time or expenses being incurred in processing these items is extreme, they should be charged as a direct project cost and not as a markup or overhead.

There are, however, situations where a markup is justified. A firm's opportunity cost of money tied up in paying for reimbursables prior to payment by the client is a justified charge. For example, if a firm could earn 2 percent interest on money in an investment account, but instead must use it to pay reimbursable vendor bills, or jeopardize its own credit standing, there is an opportunity cost incurred. A firm also incurs a real cost when it must borrow money to pay for reimbursable vendor bills for a slow paying client. When clients are willing to provide a retainer or other guarantee of prompt payment, the markup can be eliminated.

Lastly, an argument could be made for charging a markup on an item to avoid raising the firm's cost structure and penalizing all of your clients. For example, a particularly large project may significantly increase your professional liability premium. This may result in raising your overhead rate, thus making your pricing more expensive to all of your current and potential clients. By directly charging the responsible client for this cost, perhaps by treating it as a reimbursable item, you avoid this problem.

In cases where clients, in the negotiation process, refuse to allow a sufficient overhead rate to cover actual costs, designers are certainly justified in seeking a markup to cover the clearly identifiable costs of processing reimbursables. It is most important, however, that the firm's negotiators completely understand their cost structure. Markups should not be established simply because competitors are doing so or because an unaware client permits it. Markups must be justified to be valid.

This book has free material available for download from the
Web Added Value™ resource center at *www.jrosspub.com*

7

MANAGING THE DESIGN PROCESS

CASE STUDY: SELECTING THE RIGHT TEAM

Architecture would be a wonderful profession if only we didn't have to deal with clients. This is nearly a verbatim quote from one of America's leading design architects who is also a professor at a major architecture school. I heard this individual publicly make this statement at an American Institute of Architects (AIA) national convention a number of years ago. Some listeners may have been shocked by his arrogance; others thought he was only jesting; but most nodded their heads in agreement. Are all designers as elitist, arrogant, and patronizing as this individual? Of course not; however, many are perceived as such and are often considered difficult to work with. Not only do clients experience difficulty in working with designers, so do many project managers (PMs) within the same organization.

A number of years before I heard that statement, I was the operations manager for a 100-person architectural/engineering (A/E) firm headquartered in Chicago with branches in Phoenix, Denver, and Washington, D.C. This firm had pioneered many advances in project management that are now widely accepted in architectural and engineering firms, including computerized financial and project reporting systems, early computerized specification preparation, and project organizational structures and tools. All PMs were either licensed engineers or architects and had many years of professional experience. Most had extensive design and technical backgrounds, but had adjusted well to their role in the matrix/strong PM system. By the time I joined the firm, they had had a great deal of experience working with the matrix project management approach. The firm was also widely recognized for the quality of its designs and had received numerous awards.

The senior project staff in Chicago included a core group of about six to eight PMs, two lead designers (both principals in the firm), and a number of other senior and junior designers. The technical group was led by a very experienced senior associate and staffed with a cadre of knowledgeable project architects and engineers and a number of more junior technical people. In general, the system worked well except when one of the lead designers became overly enamored with a particular project. This could occur at inconvenient times. It was frustrating when they spent an inordinate amount of time on a small project. The budget would be quickly blown and all hope of a satisfactory outcome ended. The real problems occurred when they fell in love with the idea of creating the perfect design on a large, significant project that required tight planning and management.

One project in particular stands out in my mind. The client was a large teaching hospital that needed to construct a new building for its medical school. The site was very long and narrow, and adjacent to a major expressway in Chicago's Medical Center District. Compounding the problems of the site was a major elevated line for the Chicago Transit Authority heavy rail subway/elevated system that ran directly through the property. The client also wanted a building that would allow for easy retrofit for future and yet unknown technology. Federal funding had been obtained to help in paying for construction of the new facility, but the project still had a tight budget and schedule. It was clearly a difficult project requiring a great deal of creativity from all members of the project team.

Our most experienced PM was assigned to head the team and a number of very capable technical staff members were expected to join as the project progressed. The president of the firm then made a nearly fatal mistake—he selected as lead designer the principal who often found it difficult to live within a client's program, schedule, or budget. The PM protested his selection, but to no avail. This particular lead designer was very creative and worked well on projects that had a looser budget and schedule. He was a follower of legendary architect Louis Kahn and often attempted to emulate his hero. The designer was also a believer in never committing to a final design scheme and in the Inverse Pareto rule: 80 percent of the effort gets you 20 percent of the results.

As expected, it was a difficult project, one that ended over both the design and construction budgets and behind schedule. The PM was extremely frustrated and the client often exasperated. However, the design was effective and attractive and did win several important design awards.

DESIGN THEORY IN BRIEF

Changing Ideas

Architectural and engineering design during the past 125 years has evolved significantly. During the last third of the 19th century, the development of the skyscraper became the most notable manifestation of modern architecture. Starting after the great fire of 1871, Chicago became the proving ground for new technologies and materials that permitted the construction of ever taller buildings. Architects and engineers from around the nation and world poured into Chicago to express their own ideas. Soon other cities, most notably New York, copied and adapted the tall building to their own locations. Industrial architecture and engineering similarly evolved. Henry Ford's assembly line in Detroit required the development of large, one-story, reasonably clear-span manufacturing plants. Design firms such as Albert Kahn & Associates in Detroit used reinforced concrete extensively to create functional industrial facilities. Civil works such as dams, bridges, and roadways all rapidly evolved. First in Britain and later in the United States, engineers using new materials found highly creative solutions to transportation and infrastructure needs.

During the 20th century, various architectural styles came and went. The International Style, based on the ideas of the German Bauhaus, became the most influential design philosophy after World War II. In the United States, the Lever Building in New York, the Inland Steel Building in Chicago, and a number of other buildings designed by Mies van der Rohe during the 1950s and 1960s were the forerunners of thousands of less appealing copies. By the 1970s, this style had run its course. It was criticized as boring, unimaginative, and captive to economic interests desiring to maximize financial return at the lowest possible cost. In reaction, architectural theorists rejected this approach and developed new ideas.

Architectural schools became hotbeds of an intellectual approach that writer Michael Speaks has called *an almost constitutional aversion to commerce and the marketplace, the very milieu of innovation and shaper of any future architecture* ("After Theory", Michael Speaks, *Architectural Record* magazine, June 2005, pps. 71-75). According to Speaks, many design students and faculty believed *Marxist historian Manfredo Tafuri's claim that all architecture is irredeemably corrupted by capitalism.* Speaks points out *that fantasy has finally lost its allure and all connection to the real world.* Unfortunately, many architects still fail to realize that the marketplace creates the need for architecture (and in turn, architects). Most rational individuals understand that the marketplace will always prevail, design theorists notwithstanding. Technology and change are advancing at a pace requiring designers to run as fast as they can simply to keep from dropping out of the game entirely.

EVALUATING AND SELECTING DESIGNERS

What Clients and Facilities Managers Should Consider

The goals and priorities of owners and designers are often at odds. Owners seek a physical solution to a need. This solution should be functional, cost-effective, and reasonably attractive if possible. Designers wish to express their own creativity and to impose their (they believe) superior design sense on clients. Engineers often push the envelope of new technologies and products leaving clients to wonder what future problems they will inherit from these innovations. In other cases, engineers are left to solve the problem of determining how to construct designer's concepts.

Designers are often selling-oriented, not market-oriented. This means they have a package of services they offer and seek out clients interested in buying from that package. Market-oriented firms focus on the client and marketplace needs and tailor their services and prices accordingly. Many design firms seek to have potential clients use some form of qualification-based selection (QBS) and price their services on their perceived value to the client. QBS is widely used by the U.S. federal government and was instituted to permit a focus on qualifications rather than price when selecting design services. This makes great sense when purchasing special use, unique, or complex structures for a large client. For many other projects, qualifications do not vary greatly among the firms seeking the work and QBS is unnecessarily bureaucratic. Today, many clients perceive architecture and engineering services as a commodity to be purchased at the lowest reasonable price. Designers often feel inadequately compensated for the value they provide to their clients. This goes to the heart of the selling/marketing issue. A seller may believe there is great value in what they offer, but it is the marketplace that ultimately makes the decision as to how that value is priced.

Owners/clients place value on several factors when selecting designers. Service is near the top of most surveys of owner desires when hiring designers. Many are willing to pay more for a design firm with a proven record of providing excellent service and responsiveness. These service-oriented designers are almost always good listeners and communicators who respect their client's opinions and desires. Hand in hand with this service focus is respect for an owner's deadlines and budget. Owners also want as few problems as possible during the design and construction process. While some owners may have unrealistic expectations, good designers offer their clients reasonable alternatives and honest advice.

Living Together

A good designer is one who can come up with a creative, attractive, functional design within the program, schedule, design fee, and construction budget. This is not as unrealistic as it may seem. PMs can effectively manage the design process. This begins with the recognition of what drives designers. Colin Gray and Will Hughes, in *Building Design Management*, wrote, "Many designers, often driven by their inner convictions about the way the world should be, are determined to make a statement, whether political, social, monumental or aesthetic, through their work. This is developed during the process of architectural education… This system requires architects to develop their personal design philosophy and concepts and defend them strongly in open debate… There is a danger that this custom of vociferous defense may be perceived by nonarchitects as arrogance, but it is often so strongly developed as to be very difficult to modify and adapt" (Gray and Hughes 2001).

Designers are, or believe they are, artists at heart. Often, they are impractical, unrealistic, and difficult. Architectural design is not a purely creative process in the sense of painting, sculpture, or the other visual arts. Building design is performed in response to someone else's need and must always conform, at least to a degree, to that need. Rarely are designers given free reign to create for its own sake and when they are, it is typically for a unique project like a monument (Maya Lin's Vietnam Veterans Memorial in Washington, D.C.) or for a museum (Richard Meier's Getty Museum in Los Angeles). This need to create in response to someone else's requirements can make a designer's life stressful. Clients do not always have well-thought-out programs. As a project design proceeds, new ideas, additional needs, changing economics, demanding schedules, and many other items force designers to adapt. It is a life of continual change and compromise. The creative process cannot always be neatly categorized through a budget, schedule, or a program and it is almost always a collaborative process. The great genius waiting for a blinding flash of inspiration is a myth. Gray and Hughes have defined the reality as *an evolutionary process*.

These two authors have also identified constraints on the design process. Internal constraints "are imposed through wanting to work in a particular way, or with particular materials or technologies. These constraints may limit the range of solutions or may present the client with an appealing consistency. Therefore, certain designers develop a particular style of design or approach to the design problem and their reputation is based on this. Many of the constraints are self-imposed…which, if taken to the extreme, can mean that a solution that satisfies the designer can never be achieved."

External constraints are many and varied. Figure 7.1 lists some of these constraints.

These external constraints often dictate what the structure will look like, how it will be constructed and perform, what it will cost to maintain,

Client Needs	Legal/Regulatory/Standards
• Function	• Laws
• Construction Budget	• Building Codes
• Design Budget	• Performance Standards
• Life Cycle Costs	• Ethics
• Schedule	• Local Practices
• Appearance	• Community Acceptance
• Expandability	Site issues
Materials Technology/Construction Issues	• Soil Conditions
• Cost	• Water Supply
• Maintenance	• Energy Supplies
• Function	• Seismic Issues
• Constructability	• Access
• Fabrication	• Services (e.g., transportation)
• Performance	• Utilities
• Fire protection	

Figure 7.1 Some external constraints on designers

and many other factors. A designer must balance all of these limitations with his or her own aesthetic and creative desires. Unfortunately, there is no perfect solution to each design challenge.

Gray and Hughes offer the following advice to PM's when working with designers:

- Allow designers time for reflection
- Work with designers who have relevant experience and encourage and provide support to enable them to find solutions to a problem
- Establish a framework within which the tasks and objectives are kept in focus as the design moves through its stages of development
- Provide access to the client for review and provision of more information
- Help the designer understand the full implications of a new definition of the design problem and the possible need to revise the design

To these five points, I would add the following:

- Involve the lead designer in the early stages of a project—this should be as early as possible, including during marketing
- Review all aspects of the project's program, schedule, design fee budget, and the estimated construction budget with designers
- Regularly communicate all aspects of the project's status with the designers
- PMs should not attempt to design the building for designers—allow them to do their job
- Senior management should try to match a designer's skills as closely as possible to the needs of the project—do not put the Bentley guy on the Ford project
- The PM must regularly work with designers to define deliverables at each stage of the project
- Listen to your clients—without them architecture and engineering does not exist

THE DESIGN PROCESS IN BRIEF

Parts of the Design Process

As Gray and Hughes have noted, "Design is a continual trade-off between many conflicting needs until there is a solution that enables everyone to move forward to the next aspect of the problem." They have also wisely pointed out that "Designing is a process of human interaction and…the outcome contains the interpretations, perceptions, and prejudices of the people involved." In general, the design process can be divided into four parts. PMs quarterback their team through this process and regularly meet with others as required by activity on the project.

1. **Information gathering:** Gather as much information as possible given the available time and resources. In general, you are concerned with the purpose of the structure, the client's target schedule and budget, the design image they wish to convey, who will be using the facility and for what purpose, your client's decision-making process, and dozens of additional items. This information should be shared with all of those whom you believe will need it as the project proceeds. The process of information gathering continues throughout the project.
2. **Analysis:** Examine the collected data for relevance and accuracy. Based on a designer's own experience, compare the data to past projects. Question assumptions and share data with other team members. Formulate your and the team's approach to solving the client's problem. Data will continually be collected and the analysis process continues even after the completion of the building.
3. **Execution:** Undertake the preparation of design(s) and documents. Communicate continually with all team members, especially the client. Evaluate and rework decisions and documents throughout the process. Provide well-conceived and complete bid documents including drawings, specifications, reports, and analysis.
4. **Feedback:** The construction phase of a project will generate at least three-quarters of the total information on a job. This is where any unresolved problems, questions, and issues become apparent. The entire team must be responsive and react quickly to activity on the construction site. Once the project is complete, gather as much information as possible on what went well and what didn't.

Engineering Design

Engineering design is typically a great deal more technically focused than architectural design. Many elements used in engineering design such as beam sizes, chiller sizes, and electrical equipment are relatively standardized. However, to incorporate these elements into a structure requires a substantial amount of planning, analysis, and calculation. This typically includes the required structural, electrical, mechanical, plumbing, waste, and other systems. Much of a building's mechanical equipment, for example, is designed by manufacturers and requires a knowledgeable engineer to select the appropriate item for the application. As a result, it is often the engineer who actually designs the nuts and bolts of the buildings' interiors. While the architect designs the space layout and aesthetics, the engineers make it work.

Occasionally, a design architect will create special challenges for engineers. Architect Frank Gehry is famous for unusual building designs. Most of his work are one-of-a-kind projects often requiring innovative structural and mechanical design. Fortunately, modern technology, materials, and systems will allow for construction of much of this highly unusual design if a client is willing to spend enough. Some clients are more than willing to make the investment, especially if they view unusual design as a marketing device for their building, or if they are undertaking a unique building such as a museum.

Because of their focus on the structure and operation of a building (or a bridge or road), engineers tend to be much more detail focused than design architects. Unfortunately, they are often fairly far down the information line and must turn to architects, design/builders, construction managers, and other intermediaries for information. At times, they are somewhat isolated from the owner, facilities manager, or end user. This creates coordination problems and can place a burden on engineers to respond to others' needs without complete information. In some cases, such as in large civil projects, engineers work directly with the client or end user making them the primary consultant—and their role may be similar to an architect in a traditional building.

It is important for PMs in the architect's office and in the various engineering disciplines to maintain excellent communication channels to allow for timely sharing of information. This will improve the effectiveness of all organizations involved in the design and construction process.

SUMMARY

Every project begins with an idea or need. The need comes from an owner who has an expanding business, an antiquated or inefficient manufacturing plant, or has suffered a fire or flood. A growing community needs more schools, a new city hall, an expanded park, or a new art museum.

It is up to the members of the design team to convert these needs into a building, road, dam, water treatment plant, or other facility. The team needs to work together effectively with a minimum of friction and a maximum of communication. It is up to PMs in every organization involved in the project to ensure successful fulfillment of an owner's need.

8

PROJECT PHASES AND PERSONNEL RESPONSIBILITIES

PROJECT PHASES

Introduction

The typical design/bid/build project passes through five basic phases (see Figure 8.1). Phase 1 is the feasibility/programming/schematic design phase. Basic decisions regarding the building or other facility (bridge, road, etc.) are made during this phase—such as the selection of a site, the feasibility of the project, and financing. Out of these basic decisions a program is developed that details the specific building size requirements, functions of various spaces required, special equipment, and so on. Often, it is during this stage of the process that the function and design intent of the structure may dictate materials, costs, finishes, systems, and timing. Sometimes, outline specifications are prepared to more accurately detail costs. The program may also describe interior spaces in terms of requirements such as furniture and specialized equipment.

During schematic design, the general appearance of the building and the major systems (such as mechanical or structural) are determined. The need for specific products is determined at this point, while others, such as finish hardware, cannot be. Following the owner's or facility manager's

	Design/Bid/Build Project
PHASE ONE	Feasibility
	Programming
	Schematic Design
PHASE TWO	Design Development
PHASE THREE	Construction Documents
	(Working Drawings)
	(Project Manual)
PHASE FOUR	Bidding and Negotiating
PHASE FIVE	Construction Administration (Observation)

Figure 8.1 Building project phases

approval of the basic schematic design drawings, the design development (Phase 2) begins. During this phase, detailed plans for every part of the building including mechanical and electrical systems, structural systems, and materials as well as architectural details are designed. Generally, an estimated statement of construction cost based upon the design is also provided to the owner at this point.

Upon completion of the design documents, they are submitted for the client's approval. Based upon the design development drawings (Phase 3), a set of detailed, finished construction documents (drawings) are produced, schedules are developed, and complete specifications are written. These drawings and specifications become a part of the legal contract documents. The architect or other consultant, such as an engineer, also assists the owner in the preparation of all necessary materials for the bidding process. This generally includes bidding information, bidding forms, conditions of the contract, and owner-contractor agreements.

Following the owner's or facility manager's approval of the construction documents, the consultant assists the owner in obtaining bids or negotiated proposals and assists in the awarding of the contracts (bidding and negotiation Phase 4). The bidding process in private work can be on a closed or selective bid basis where the owner asks for negotiated bids from several contractors on a preselected basis. Some private owners use an open bidding process where all interested parties are asked to bid and the contract is awarded on a low-bid basis. In public work, the bidding generally is open to all contractors who meet the bonding requirements.

Normally, it is the architect who receives the bids, presents them to the owner, and includes any recommendations. If the owner has retained a construction manager (CM), the CM may receive the bids. The construction documents and specifications become part of the contract and are legally binding agreements that may not be changed without the owner's signed approval. Usually, any changes flow through the architect for his or her review. Subcontractors are hired by the contractor or CM, and subcontracts are binding in the same manner as the general contract.

Construction administration (Phase 5) begins with the awarding of the contract. Generally (except when a CM is used), the architect is the owner's representative during the construction process. However, he or she is there only as an observer and has no right to stop the work or to authorize changes except those that are approved by the owner. As an observer, it is the responsibility of the architect to keep the owner informed as to the quality and progress of the work. The architect does have the option of rejecting work that does not conform to the contract documents. He or she is also responsible for reviewing and approving shop drawings, product data, and samples, but only for conformance to the contract documents.

The architect processes the contractors' applications for payment and issues a certification for payment when the construction work has proceeded (based on his or her observations) to a point indicated in the contract where the contractor is entitled to a progress payment. (For those who both furnish and install products, payment would likely be on the same basis.) As construction proceeds, the contractor must submit manufacturers' and subcontractors' shop drawings for the architect's review. In addition, the professional can ask for additional product data or require product tests.

It is clear that the owner should be involved in every stage of the construction process. Initially it is the owner who determines the need for a facility and along with his or her consultants (which may include the design professionals) determines feasibility, estimated budget, project scope, and so on. During design development, the owner is responsible for approving the designer's drawings and reviewing the construction budget. The construction document and bidding and negotiating phases require the owner's review of the finished construction drawings and involvement in the bidding and negotiating process. It is the owner who awards the contract for construction and the architect serves as coordinator.

The construction phase requires the owner or facility manager to meet regularly with the architect to review the project's progress and problems, and review and approve change orders and certificates for payments. The owner is involved in nearly every aspect of the construction process.

DESIGN FIRM STAFF AND RESPONSIBILITIES

Position Descriptions

There are a wide variety of terms used to describe positions in a design professional's office. Though the same term is used, such as project manager (PM), the responsibilities of an individual may also vary greatly from firm to firm. The terms used in this chapter and the descriptions given are those most commonly found in professional firms, and are in accordance with the definitions used by the American Institute of Architects (AIA) and the American Council of Engineering Companies (ACEC).

The highest level of project authority in the designer's office is the principal or partner in charge (PIC). Frequently, this is the individual who was initially responsible for obtaining the contract. In large firms, he or she is rarely involved with any of the details of the project, yet often serves as the final authority (except for the client) on the project team. Usually he or she serves as liaison to ensure the project is proceeding according to the client's wishes. Some principals take a much more active role and function as the PM; however, this is more common in smaller firms.

The PM is responsible for the administration of all phases of the project, supervises project personnel, and handles project scheduling (see project flow diagram, Figure 8.2). He or she confers and corresponds with the client, outside agencies, consultants, and so on. The PM is also responsible for project records and obtaining approvals.

A project designer generally initiates and creates designs and planning of projects and participates in all phases of its development. He or she confers and corresponds with the client in conjunction with the PM to determine the project's requirements.

Job captains are responsible for planning and coordination of a project under the direction of the PM. They direct groups of drafts people who do the actual drawings. (Today, nearly all drawings are completed using computer-aided design and drafting [CADD] software.) In many firms the job captain is brought into the project in its early phases. It is the job captain who (along with the specification writer) specifies and details a product. The job captain is interested in the verification of function and layout, but his or her primary job is focused on the preparation of detailed drawings. On smaller or less complex projects, an experienced job captain may actually serve as the PM. In some firms, the job captain is called the *project architect* or *project engineer*.

Project Phases and Personnel Responsibilities 131

Project Phases

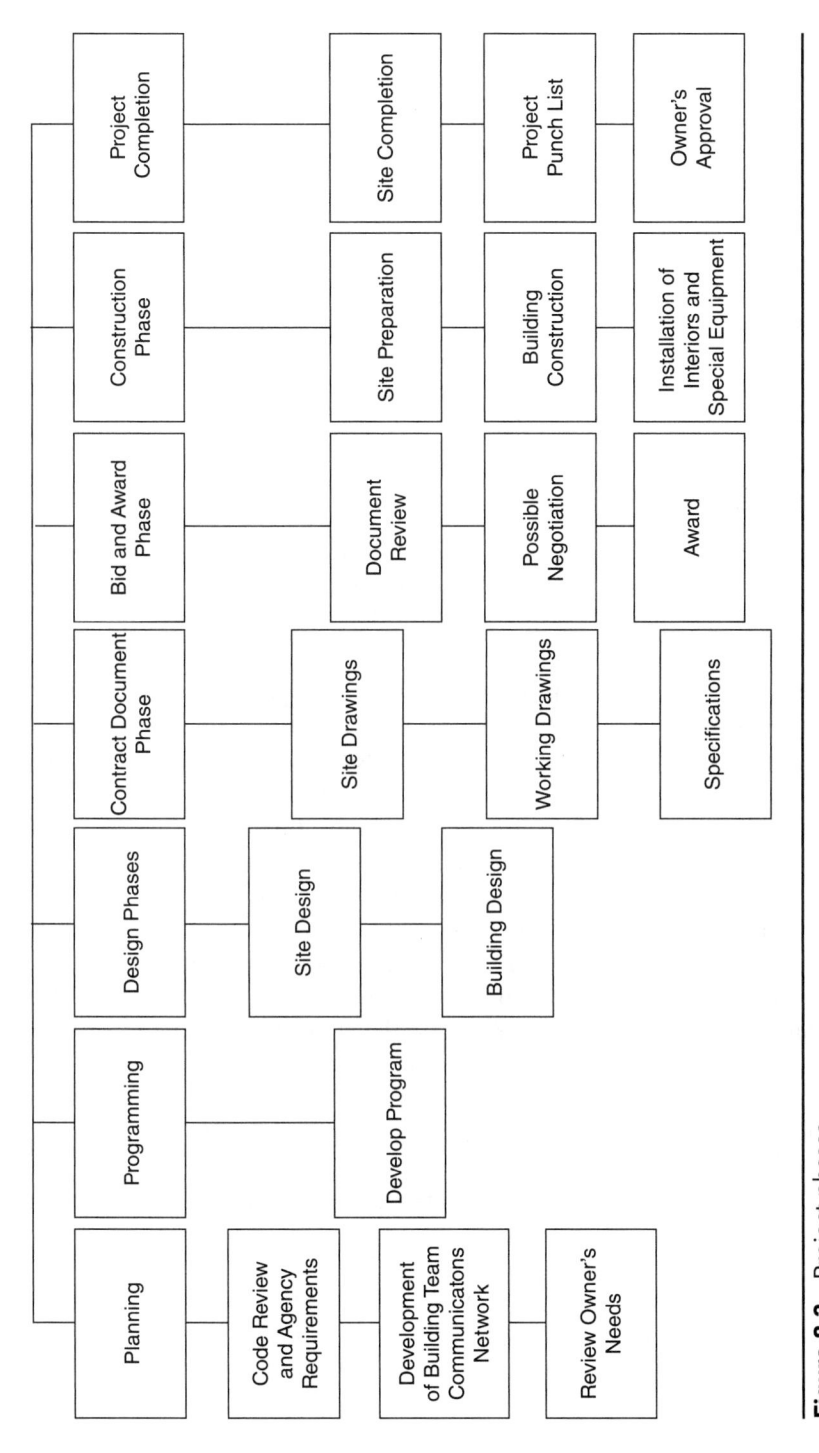

- Planning
- Programming
  - Code Review and Agency Requirements
  - Development of Building Team Communicatons Network
  - Develop Program
  - Review Owner's Needs
- Design Phases
  - Site Design
    - Site Drawings
  - Building Design
- Contract Document Phase
  - Site Drawings
  - Working Drawings
  - Specifications
- Bid and Award Phase
  - Document Review
  - Possible Negotiation
  - Award
- Construction Phase
  - Site Preparation
  - Building Construction
  - Installation of Interiors and Special Equipment
- Project Completion
  - Site Completion
  - Project Punch List
  - Owner's Approval

Figure 8.2 Project phases

Specification writers are responsible for verification of function and layout and preparation of detail drawings along with the job captain. Most of all, they are responsible for the verbal description and appropriate use of all materials and products used in the project. The construction phase requires the owner or facility manager to meet regularly with the architect to review the project's progress and problems, and review and approve change orders and certificates for payments. The owner is involved in nearly every aspect of the construction process. The field representative (rep) is responsible for the general administration of project construction contracts; periodic observation of and reporting on construction activities, construction records, and progress schedules; and reviews construction and bidding documents.

Specialized Consultants

A wide range of specialists has appeared on the construction scene in recent decades. Most are hired by clients to aid in cost control and/or provide management skills lacking in the owner's organization. How successful these specialists are is a subject of some debate. At the core of this debate is a turf dispute on the part of other traditional industry consultants, such as architects. These traditional consultants would like to preserve their role in the construction process, carve out new niches for themselves, and prevent new competitors from achieving a foothold. This issue would be of little consequence if traditional providers were meeting the needs of clients. Unfortunately, many are not.

The traditionalists claim that the specialists are simply glib marketers praying on the fear and ignorance of owners. They also claim that cost savings are often illusionary as the additional layer of team members adds to communication problems, thus increasing or shifting costs. The specialists reply that they have developed their services based on the very real needs of owners. They also point out that traditional providers do not or will not offer the services clients need. Many owners also resent the attitude of architects and others that they are simply the rubes of slick marketers.

Why have traditionalists failed to meet the real needs of clients? There are several reasons.

1. **Liability concerns head the list of many:** Risk is a significant issue in the construction industry. Standard contracts are often written to shift or share risk with other parties. Many traditional providers

are unwilling to assume risk for items such as construction costs and construction supervision.
2. **Fees are often inadequate for the level of risk required:** Traditional providers such as architects argue that they are willing to assume greater risk for projects if the fees they are paid are high enough to compensate them for the risk. To many, it is a simple risk/reward equation.
3. **Many traditional providers lack the management skill, expertise, or staff to provide the services clients require:** As a result, they are unable to embrace new services. Often, new or expanded services require training lacking in the education or experience of traditional consultants such as architects and engineers.
4. **Promoting what many traditionalists call nontraditional services requires new marketing approaches and skills that they lack:** Clients are knowledgeable and are seeking cost control, attention to service, and management skill. While design is an important issue, it is often of secondary importance to many clients.
5. **Projects are much more complicated than found in the past:** Technological change requires new expertise and services. For many buildings and projects, no single engineer or architect has all the required expertise, necessitating the use of specialists.

Specialists and Generalists

Who are some of these new (and not so new) specialists and generalists? They can be divided into two groups: *project specific* and *generalists*. The project specific consultants include technology experts such as computer specialists, commissioning agents, and value engineers. Several of these types of project specific consultants will be discussed in later chapters of this book. Some of the types of generalists are discussed here:

1. **Project management consultants:** These are firms or individuals who specialize in managing a particular project for a client who lacks the skill, staff, time, or interest to do so. For example, a hospital adding a doctors' office building might turn to a project management consultant to help define the scope, determine funding needs and devices, hire architects and engineers, and so on. A variation of this role is used by an increasing number of corporations and institutions who do not wish to hire their own employee PMs or facility managers. These organizations are outsourcing their

work to contract project management firms who employ individuals to work on the corporation's or institute's projects. It gives corporations and institutions great flexibility in handling work and does not lock them into commitments to employees, the payment of fringe benefits, and the like.
2. **Program management consultants:** These consultants are similar to project management consultants except that they may manage an entire building program. A school district undertaking a replacement program or adding a number of new schools, undertaking renovations or life safety work, etc., may hire this type of consultant to supplement its own staff or to provide the necessary expertise. Typically, these projects are undertaken over a number of years and in multiple locations. The client may lack sufficient staff to manage this work, but can effectively manage the work of the program management firm.
3. **Design/build firms:** These types of firms are used by owners seeking one-stop shopping for both the design and construction. The belief is that communication and cost control are improved when both activities occur within the same organization. A corporation requiring a new manufacturing facility may find this a very effective approach. Design/build services has had tremendous growth in recent years and is now the delivery method of choice for many owners. A caution is necessary; however, as many design/build firms are actually only a combination of two independent organizations, allied to seek a project. This may call some of the possible benefits of the approach into question.
4. **Construction managers:** These types of managers typically justify their fee by the cost savings they can identify during the design and construction process. This may occur based on a product substitution, alternative erection process, value engineering, or other methods. CMs can be very effective for large, complex projects such as a significant public building, highway reconstruction, or other similar jobs. Most CMs are commonly at-risk for the cost of construction. Some are not-at-risk and are hired simply for their expertise in managing construction projects or because of inadequate client staff. On not-at-risk projects, firms are typically paid a fee and are not exposed to loss or benefit based upon the construction cost coming in above or below budget/expectation. In this case, he or she typically bills on an hourly basis for his or her time. I worked with one state agency that used this approach

to supplement their own staff when workloads were high. It was expensive in the short term, but avoided permanent hiring.
5. **Architects:** Architectural consultants are particularly effective in traditional design/bid/build situations where there is sufficient time for the structured process the name implies. Time and money must be allowed for design changes as the process unfolds.
6. **Engineers:** Engineering consultants have a multitude of roles. Some are lead consultants on projects that are heavily civil in nature, or may be primarily a mechanical engineering design project. Often, engineers serve as sub-consultants to others and are down the food chain for fees, payments, and communication with the client.

DESIGNER/CLIENT RELATIONSHIPS

Client Retention

It is crucial for design firms to provide high quality service to keep their clients satisfied. Many design practices obtain the largest part of their work from repeat clients. One financial industry observer has identified five keys to client retention:

1. **Respect:** Treat clients, whether institutional or individual, with respect. If you respect your clients, they will in turn respect you and your organization. Take the time to find out what they want, listen to what they have to say, and, give them what they need, not what you assume or think they want.
2. **Communication:** Communication goes hand in hand with respect. Keeping everyone well-informed makes the team effort a reality. This includes all aspects of written and oral communication. From timely invoicing, monthly reports, periodic telephone calls, and personal visits, everything must be complete and accurate. Review your clients' status on a regular basis, contact them and ask for suggestions and comments, and determine if they need anything or if there are problem areas.
3. **Service:** Emphasize to your staff that they must offer prompt and complete service to clients. Research and implement controls and systems to ensure that clients are getting the best service you can provide—this ranges from telephone procedures to individualized attention from staff. Quick responses to requests and immediate

resolutions to problems are critical. Respond to telephone calls in a personalized manner with courtesy and intelligence.
4. **Satisfaction:** A satisfied client translates into loyalty and referrals. The combination of respect, communication, and service make a client remain with you. Satisfaction also means coping with problems in a way that all parties achieve their objectives and you maintain your client relationships. Communicate to your staff the necessity of quick resolutions of problems. A problem should never become an issue, at which time it may be too late to avoid damaging a relationship.
5. **Loyalty:** Stand by your clients. Seek ways to offer your help, resources, and abilities. Do your best to help solve their problems and needs within their budget and time frame. Your efforts will be repaid with both client loyalty and a positive image that is communicated to the entire industry.

Owner/Client Expectations

It is surprising how many design professionals fail to fully understand their client's needs and wants. Many designers have little interest in learning about client priorities, methods of operation, or information needs. Some professionals are even condescending to clients. These designers have the attitude that they know what is best for the client and want no interference. Unfortunately, these designers forget whose project it actually is. Every owner/client is different, and their needs and wants also vary. There are, however, some needs that are common to most owners.

Owners and designers are often moving in opposite directions on this subject. In recent years, there has been a trend toward encouraging owners to hire all consultants directly. This is due in large measure to liability concerns. Designers can also be found liable for the errors and omissions of consultants if they hire them. When a client hires a consultant, the prime designer's liability is lessened. On the other hand, owners are looking for one party to be responsible and in charge of the project. As projects have become increasingly complex, this need has grown. The result has been the development of the various specialists to fill the gap left by engineers and architects. These project management and construction management firms are now assuming the responsibilities formerly held by many design firms. The result is deterioration in the designer's role, scope of services, and authority.

A design firm's internal project management structure can also frustrate many owners/clients. Owners are looking for one individual to be in charge and responsible for their project. When they have a question or problem, they must know whom to contact. Unfortunately, many design firms operate on a crisis management basis. Often, especially in smaller design firms, principals try to run projects, bring in new work, and also manage the firm. Rarely do they succeed in all of these activities. A departmentally organized firm can only make the situation worse. In these firms, a series of changing individuals are responsible for the project during its various phases. As a result, there is no consistent point of contact. Most knowledgeable owners favor the matrix or strong project management system of project delivery. In this system, the client knows who is in charge of the project and whom to contact. Most all experienced and sophisticated clients endorse this approach and many employ it in their own organizations.

Specifically, clients/owners want the following:

1. **Clients want to be kept informed:** It is a common complaint among owners that their design firms fail to keep them adequately informed on the project's progress and the design options involved. Designers must develop reporting systems, meeting processes, and monitoring tools to keep clients informed.
2. **Clients want good cost control:** To some owners, the designers seem to be unaware of the costs of design decisions. To many designers, some owners want more out of a facility than they are willing to pay for. While this dichotomy in goals may never be fully resolved, firms that show a concern for construction costs at an early stage of the project are most appreciated by owners. This situation is particularly acute for public sector clients. Many of these agencies receive their funding based upon an appropriation and find it very difficult to obtain money to pay for increased construction costs.
3. **Clients want technical competence:** Although many designers don't believe it, most owners are more concerned with technical competence and experience than they are with the designer's fee. Clearly, design fees are an issue to many clients. However, most owners repeatedly indicate that fees are not as important an issue as many designers believe.
4. **Clients also want to be invoiced regularly:** This is essential if they are to plan and manage their own cash flow. Many designers are

slow at billing or fail to provide complete and accurate information. The result is a delay in processing invoices and potential irritation to all parties. In a well-managed design firm, it is the PM who obtains information on billing needs and requirements at the start of a project.

What Owners Should Ask Designers before Selecting Them

Several of the design professional societies publish guidelines for owners to use in evaluating and selecting engineers and architects. Most of these publications tend to be slanted toward the traditional design/bid/build delivery methods, fail to ask crucial questions, or attempt to encourage the potential client to hire the particular type of professional the publication was prepared by. Every PM must be prepared to provide a potential or current client with a wide range of information. Failure to completely or honestly provide the requested information may cost a design firm the contract. Some items a sophisticated client may ask a design firm include the following:

1. **Experience:** Of particular importance is a design firm's experience with similar projects.
2. **Pricing justification:** The design firm must justify its pricing scheme including overhead rates, profit targets, and overall multipliers.
3. **Explanation of the project delivery system:** The most successful designer/owner relationships occur when both have similar or complementary systems. For example, the design firm may have a single-discipline departmentally organized firm and the client has a sophisticated project management system. Under this circumstance, it may be difficult to work together.
4. **Specific team members:** Some experienced clients contractually require that the PM and other key team members be named in the contract. The design firm is not allowed to change team members without the specific approval of the client. (Some designers wish they could do the same to clients who often change their own PMs or facilities staff.)
5. **Fee:** Although most designers would like to use a qualification-based selection (QBS) process, fee is often a factor. During negotiations, it is vital that the design firm PM fully understand

their firm's pricing scheme, proposed budget (and underlying assumptions), proposed scope of services, change order management process, and any other significant items.
6. **Project management manual:** Experienced clients sometimes request a copy of a design firm's project management manual. This manual documents a firm's procedures and can help to reassure a potential client that you have well-thought-out processes, systems, forms, and so on. This also gives clients the opportunity to understand how your firm operates and how you will work together on the project. Unfortunately, most design firms do not have a written (partial or complete) project management manual. Frequently, something is simply thrown together when a request for proposal (RFP) is received from a potential client. This type of manual will be discussed further in Chapter 9.

Successful Projects Require Assertive and Knowledgeable Owners and Facilities Managers

Many construction projects suffer unnecessarily from schedule delays, cost overruns, conflicts, and poor performance by team members. Finger pointing and passing blame often become a major focus of effort for the parties. At the center of this storm is the owner, who has retained all of the other parties to solve a need and is paying for the problems that have arisen. No matter how much effort is spent, some problems cannot be avoided. However, it is certainly to the owner's advantage to have a smooth running project designed and constructed by a team with good communications and a good working relationship.

Some projects are fated to be over budget and schedule because of the owner/client's failure to be assertive and knowledgeable. What should an owner do to ensure a smooth-running project? The following are some key items that every owner should consider:

1. **The owner's organization must have a facilities management PM to represent their needs:** Typically, this is an individual in the owner's employment (or is an outside contract staffer such as a facilities management firm) who acts as the owner's representative to all other team members. This person must have extensive experience in managing similar projects and, most importantly, have authority commensurate with his or her responsibilities. Many owners' PMs lack this authority and simply act as intermediaries.

For example, in most government agencies, PMs cannot authorize additional expenditures for design or construction changes without the approval of senior managers or bureaucrats. As a result, needed changes are often extensively delayed. In many owner/client organizations, PMs are not brought into the process until many decisions have been made regarding the form, function, target cost, and schedule for constructing a facility. Their earlier input could have saved much wasted time and expense, and potentially avoided mistakes.

2. **Often, owners' PMs are weak in project management experience, lack authority, and have poor communication skills:** Frequently, owners' PMs are hired or selected because of their technical ability, not their project management skills. In some cases, they are hired from non-construction backgrounds and have little experience in their new assignment. Successful projects depend greatly on an owner's rep who knows what he or she is doing and is assertive.

3. **The owner's PM must assume an active role throughout the project:** Outside specialists such as CMs, engineers, architects, and contractors are hired for their particular expertise. Nearly all are well-intended, competent, and conscientious professionals who will perform to the best of their ability. None of them, however, can completely represent the owner as each has their own biases, experiences, and profit motives. Only the owner's PM can make the final decision regarding the project's scope, schedule, and budget. This is a responsibility that cannot be delegated and requires involvement from start to finish on the project.

4. **The owner and the owner's PM must have a good understanding of the appropriate role of each team member in the design and construction process:** The construction industry includes many parties who are marketing their particular service. Not all services offered by generalists and specialists are appropriate for all circumstances and all projects. The roles of some of these parties were discussed earlier in this chapter.

5. **Owners must understand the various project delivery methods available to them:** These methods will vary depending upon the nature of the project, the size of the project, the time frame available until the project is needed, the project's funding mechanism, and many other factors. Although these methods were discussed in Chapter 1, additional information is provided below.

a) **Design/Bid/Build:** This system is considered the traditional method of project delivery. Under this method, design is completed prior to bidding. Bids are submitted by one or more general contractors for the work and include the subcontractors. A bid is accepted or negotiations occur based upon suggested changes in the specifications, such as a proposed product substitution. Construction, subject to change orders, occurs and is completed based upon an established scope, schedule, and price.

b) **Design/Build:** This approach has become a much more widespread delivery method in recent decades. Under this method, a design/build firm or team submits a proposed scope and fee for both the architectural/engineering design and the construction of the project. This single point of responsibility holds much appeal for many owners as they have only one party (the design/builder) to hold accountable and only one party at which to direct project communications. Design/Build is even being used in projects, such as highways, that would not seem to require this approach. Critics claims that design/build leads to less creative and attractive design. Proponents claim that the relative turnkey nature of design/build greatly benefits owners and outweighs any aesthetic concerns.

c) **Fast-track:** This method of design and construction is very useful for extremely time-sensitive projects where the opportunity cost of not having a facility far outweighs the added cost of design and construction. For example, a manufacturer not having sufficient production capacity will potentially lose more in sales and profits than the cost of fast-tracking a construction project to gain more capacity. Fast-track projects require breaking the work into bid packages prior to beginning design and ordering long lead time items (such as mechanical systems) as early as possible based upon realistic estimates of final need. Owner involvement and coordination are crucial for the success of a fast-track project.

6. **Owners must have a good understanding of project team organizational systems:** Owners must also select team members based upon the effectiveness of these systems. For example, as noted earlier, an owner organization with strong PMs possessing both

authority and responsibility will likely be very frustrated working with a departmentally organized firm. Many of these firms lack PMs and may have four or five department heads each in charge of only part of the project. (See Chapter 1 for a detailed discussion of project team organizational systems.)
7. **Owners must understand how to evaluate proposals from engineers and architects:** Price is an important issue, but should not be at the top of an owner's criteria for evaluating proposals. As a general comment, I would be very willing to select a design consultant who is a little less experienced, slightly less technically capable, or a bit less creative in exchange for one who provides top-notch service and is strong on communication skills.
8. **Owners must have effective, well-thought-out project management procedures of their own:** A project management manual would be very helpful for an owner's internal organization. It is a valuable document to share with outside consultants and contractors. The manual typically documents procedures, forms, requirements, and other items and will improve performance by allowing everyone to understand the owner's operations and expectations. The owner should review this manual with all design and construction consultants before beginning work. This early stage meeting should set the groundwork for future communications and working relationships. Obviously, if a formal Partnering process is being employed, then the initial Partnering meeting would be an appropriate forum for reviewing procedures and forms. (Partnering will be discussed in Chapter 9.)
9. **New technology:** Many items of new technology have been adopted by the industry only after major owner/clients such as the U.S. Army Corps of Engineers have required their use. CADD use experienced its greatest growth after government agencies required submittal of computer-aided drawings. The use of Building Information Modeling (BIM) software is now experiencing similar growth. This software is a tool for improving project coordination and communication—and will be discussed in Chapter 13.
10. **Project programs are increasingly being provided to design consultants by the client:** Where the project type is one often constructed by the owner, these programs can be very complete and detailed. In some cases, an owner may retain the services of a specialist firm to develop the initial program. In other situations, the prime design firm will complete a program before undertaking

design. However this is accomplished, success is dependent on the completeness of the information provided, the timeliness of the data, and of the ability of the owner to communicate his or her needs and wishes.

11. **Owners must have realistic construction budgets and project expectations:** One of the most common afflictions impacting both designers and owners are unrealistic expectations for the cost and performance of projects. Many owners do not understand the costs of construction or the impact that seemingly minor changes can have on the overall project cost. They must be willing to pay designers for the *what ifs* and the changes in scope they require. Designers must realize they need to meet budget, schedule, and program requirements. If high construction bids require redesign, then owners must be willing to pay at least some of the cost of redesign.

    Unrealistic building performance expectations are common among even experienced clients. Every building, no matter how much time and money is spent, will have flaws. Someone is bound to be cold, hot, feel a draft, need more light, or have other preferences. The building should not have leaks, create serious health hazards, or have structural failures. There is however, a generally accepted standard of performance set by codes, standards, and laws that should be the minimum acceptable level of performance. Owners and designers should talk about their expectations before progressing far on the project.

12. **Local practices and conditions:** Owners and designers should learn about the local practices and conditions where they are planning a project. Problems can arise on projects when the owner and designers are unaware of the local conditions under which the design team and contractors must work. Unique codes, requirements, labor conditions, and practices can make a project much more difficult than was experienced by the owner in the past or at other locations.

    In my former neighborhood in Chicago, a large, Seattle-based, national warehouse retailer wanted to construct a store on a vacated industrial site. They hired their usual Seattle architect and engineer to design the facility. Chicago does not follow (and never has followed) the Uniform Building Code (UBC). The city has a unique, difficult, and convoluted code designed to give maximum flexibility to local aldermen and politicians. The owner and

designer expected to come into Chicago and construct a store much as they have hundreds of times in suburban locations. They quickly learned a hard lesson in big city politics. The store was constructed, but after years of delays and at a much higher cost.
13. **Owners should not force the design team to bid drawings prematurely:** If the owner and designers do not agree that the drawings are ready for bidding, they should not release them. There have been many well-publicized cases of drawings being bid before they were complete. Usually, this happens on projects where there have been many design change orders and coordination may have been lacking. The University of Washington case discussed in Chapter 1 illustrates the consequences of premature bidding.
14. **Owners need to be aware of the complexity of current mechanical systems and plan for operations and maintenance (O&M):** Mechanical engineers rarely are involved in O&M, but often take the blame for poor system performance. Mechanical subcontractors supply and install the equipment and usually come back and test and balance the equipment before the one-year warranty period expires. After that time, the owner is often on their own. Many fail to plan for the ongoing O&M of the systems, training of staff, documentation of maintenance, storage of manuals and supplies, and so on.
15. **Owners should encourage a Partnering process:** This process is often used for civil engineering work, particularly by departments of transportation.

Summary

While no panacea, an assertive, knowledgeable owner can improve the odds for a relatively trouble-free design and construction project. Without addressing the issues raised in this article, your project has almost no chance of warding off problems and conflicts.

OWNER PROGRAM MANAGEMENT OPTIONS*

Owners considering a construction project have a wide variety of options for delivering design and construction services. The owner must secure or provide planning and design services, purchase the goods and services required to build the project, and perform effective management

---

* The material in this section was prepared by Richard Pearce to offer an additional perspective on program management.

throughout the project. Many delivery options are possible and no single approach is best. Each option places major participants in different roles and assigns different responsibilities. Each allocates risk—and reward—in a different way. Depending on the situation and the owner's needs, each is more or less appropriate. Regardless of the form of delivery option, owner decisions and actions in critical areas has its greatest impact on other participants in the project and on the success or failure of the owner's project.

Issues in Selecting Delivery Options

The choice of delivery approach is not a casual one—for the owner, architect, and contractor. Owners are advised to employ approaches that respond to their project types, priorities, management capabilities, approach to risk, and desire for involvement in design and construction decisions. Several factors in the owner's management capabilities are critical to a successful project:

- Experience, clarity, and flexibility in establishing project parameters
- Approach to making decisions and resolving differences
- Willingness to compensate design professionals fairly for the risks they are asked to assume

Owner Requirements

Within the scope of the project, the owner seeks three classic design goals: (1) highest quality, (2) lowest cost, and (3) shortest time. Everybody wants all three. However, priorities may vary with many owners faced with fixed budgets, nonnegotiable deadlines, or inflexible quality standards. The owner's involvement is a key factor. Some owners desire hands-on control of key decisions. Others look to the building enterprise to select and qualify sites, finance the project, or even own and operate it for an extended period.

Owner Capabilities

Against the backdrop of owner needs lies the issue of owner capabilities. What resources or experience does the owner have for project management, design, and/or construction? What is the owner's ability to make decisions and can he or she accept the consequences of those decisions?

Owners considering a building project have a wide variety of options for delivering design and construction services. The owner must secure or provide planning and design services, purchase the goods and services required to build the project, and perform effective management throughout the project. Many delivery options are possible and no single approach is best. Each option places major participants in different roles and assigns different responsibilities. Each allocates risk and reward in a different way. Depending on the situation and the owner's needs—each is more or less appropriate. Regardless of the form of delivery option, owner decisions and actions in critical areas have their greatest impact on other participants in the project and on the success or failure of the owner's project.

Owner Responsibilities

The conventional allocation of owner responsibilities in building projects consists of the following:

- Project definition and site
- Surveys and soil tests
- Document review
- Property insurance

Where an owner has in-house design and construction management capabilities, owner involvement may include, in addition to the responsibilities just mentioned, the following: programming, dispute resolution, change orders, certification of payments for subcontractors and purchasing, bonds, cost estimates, bidding, submittals, inspections, permits, government approvals, general conditions, guaranteed maximum price, safety and general conditions, and communications, on-site coordination, and scheduling. The options, therefore, are dependent on the owner's desire for involvement. Owners who desire hands-on management of their projects and who are prepared to build their own staff capabilities will manage the program or project themselves. Owners without in-house management capabilities will contract out these activities and, in effect, outsource all services required for their project.

Delivery Options

Owners have a wide spectrum of options for procuring design and construction services. The owner may provide services with his or her own in-house staff or contract for both design and construction services through a combination of distinctive project delivery approaches. Many owners utilize the traditional design/bid/build approach, construction management and design/build approaches, in a variety of combinations that provide delivery options responsive to owner needs. Some owners—especially large commercial, institutional, and public clients—are constantly building and rebuilding. These owners may see projects as components of larger facilities design and construction programs—programs to renovate existing complexes, upgrade for energy efficiency, remedy hazards, or incorporate new office technologies.

These owners may structure a delivery approach at the program level (all projects in this program will be delivered in the same way), or even procure design and/or construction services for the entire program—leaving it to the vendor to decide how best to deliver each project. Looking at program delivery, owners may choose the following:

- Manage the program themselves, selecting the appropriate delivery option and procuring design and construction services for each individual project
- Contract with an outside program manager who advises on the selection of delivery options and procurement of design and construction

These options lie along a continuum (see Figure 8.3). Moving to the left, the owner's role shifts from managing the program to managing the program manager. Owners who desire hands-on management of their projects and who are prepared to build their own staff capabilities will stay at the right. Owners who prefer to contract out these activities, who face position freezes (and cannot develop in-house design and construction management capabilities), or who are strategically outsourcing services once provided in-house will move to the left.

Program Management: Extension of the Owner's Staff

Program management is not a delivery option. It is best described as doing for owners what the owner would do if they had their own in-house capabilities. Program management is a term applied to the process involved when an owner undertakes a building project. The owner must secure planning and design services, purchase the goods and services needed to build the project, and ensure effective management. Selecting

Design/Bid/Build

```
                        Owner
          ┌───────────────┼───────────────┐
   Program Manager ─── Architect/Engineer ─── General Contractor
                                                      │
                                                 Subcontractor
```

Design/Build

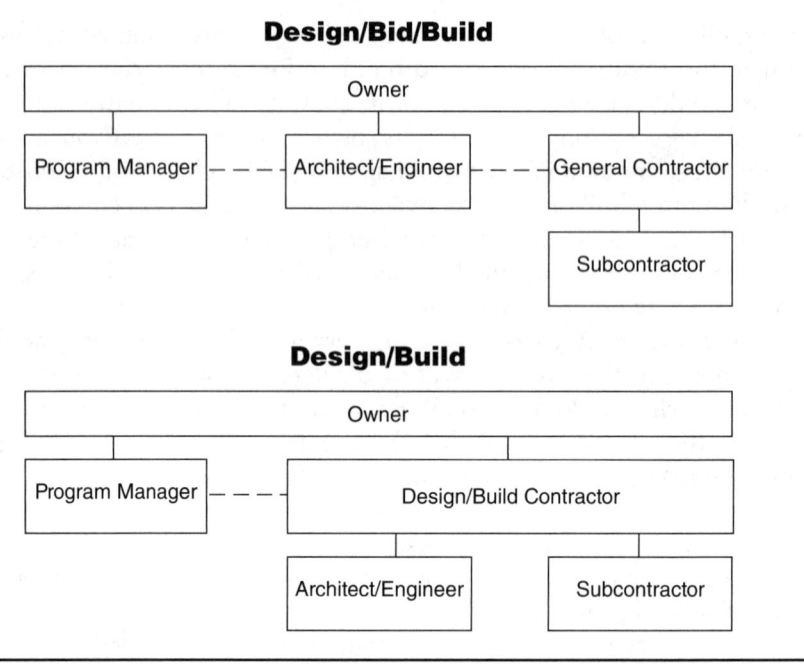

```
                        Owner
          ┌───────────────┴───────────────┐
   Program Manager ─────────── Design/Build Contractor
                                 ┌────────┴────────┐
                          Architect/Engineer   Subcontractor
```

Figure 8.3 Contractual relationships

the appropriate delivery option and procuring design and construction services may be a management capability that some owners have within their own organizations. For many owners, however, the issue of available owner management resources for design, project management, and/or construction responsibilities is not realistic and is often inconsistent and in conflict with their primary business. For these owners, many building professionals offer program management as a consulting service.

The successful program management consulting service is seen as operating as an extension of the owner's staff. Owners, therefore, can strategically outsource services once provided in-house. The more compelling argument is that the owner can stick to his or her core business (e.g., educating students, manufacturing, social services) and rely on outside building industry professionals.

Program Management: Scope of Services

The scope of services may include pre-project planning, programming, design, construction, occupancy, and facilities management. The contract is an agent/program manager relationship, with all design and construction contracts signed directly by the owner. For example, the owner contracts with an outside PM who advises on the selection of delivery options and procurement of design and construction. Actual project delivery is accomplished under a variety of approaches (e.g., design/bid/build or design/build) for any project or site with program management responsibility.

Summary: Program Management Applicability

- Program management allows an owner the resources to concentrate on their core business with building professionals serving as an extension of the owner's staff, and representing and acting as agent in the design and construction process.
- Program management incorporates team building to improve communications—working together to anticipate, identify, and solve problems as quickly as possible.
- The PM serving as an adjunct to the owner's organization provides the owner with the degree of construction sophistication that many owners do not have internally.
- Program management services are most frequently employed by occasional building owners. These owners are often unfamiliar or limited in their experience with the construction process.
- Program management services are especially appropriate for developing construction programs to renovate existing complexes, upgrade facilities, remedy hazards or problems, and incorporate new technologies.
- Program management's flexibility allows a variety of delivery options (e.g., design/bid/build or design/build) for each discrete project or site with program management oversight.
- Program management identifies and manages the owner's risks in the process of developing, designing, and constructing his or her building project.

This book has free material available for download from the Web Added Value™ resource center at *www.jrosspub.com*

9

CONTRACT MANAGEMENT/PROJECT ADMINISTRATION

BEFORE SIGNING THE DESIGN SERVICES CONTRACT

General Comments

A contract is the baseline understanding for the project and the project manager (PM) should be heavily involved in determining the contents of this document. The material that follows is intended to provide designers with key items to consider *before* negotiating a contract with a (potential) owner/client. Also, owners and their facilities managers must be aware of how designers approach contracts and the consequences of poorly prepared contracts on their projects. While reliance on standard form contracts prepared by the Engineers Joint Contract Documents Committee (EJCDC), the American Institute of Architects (AIA), and others is commonplace, many designers are poorly prepared to complete these contracts.

Some factors for engineers and architects to consider before signing a design services contract include:

- Evaluate your marketing objectives/potential client targets

- Consider your ability to provide the services that your potential clients require
- Evaluate your experiences in working with similar clients
- Specialization may enhance performance/cost control through knowledge gained by your staff in working on similar projects in the past
- You may choose to decline (or not seek) a project based upon past experience
- Inexperience with the project or client type, current/projected workloads, etc., may influence your decision

FINANCIAL ISSUES

1. **A warning:** Seeking and obtaining design work for which a design firm is not well suited can be a costly experience both financially and emotionally, and may damage the firm's reputation. For example, a design firm that has never worked for a federal agency needs to learn a great deal about the federal contracting process. A firm may be technically qualified, but not well versed in the financial issues involved in federal contacting. (See the *Federal Design Work* case in Chapter 6.) For the PM, this can require a costly education. To successfully evaluate a design firm's ability to complete a contract requires understanding a number of financial issues, including:
   a. Knowing your own costs and cost structure
   b. Knowing your own cost of doing business such as multipliers/overhead rates/profit targets
   c. Knowing your own costs allows a PM and firm to determine if they can make money doing the project or provide the necessary level of service

   No sensible manufacturer of a product would produce and offer an item for sale without knowing the full cost of the component parts, or the costs of assembly, marketing, overhead, distribution, service, etc. Unfortunately, many engineering and architectural firms regularly offer services without an understanding of their cost of doing business as reflected in their project multiplier. As a result, no matter the form of the fee quote to a client (fixed fee, percent of construction, lump sum, square footage, time card, etc.), the PM may lack confidence that the underlying information they are using for project budgeting is accurate. The result

can be a project that meets the job labor budget, but still does not achieve target profits.
2. **Scope of services:** Before a contract is signed, a well written scope of services defining the work to be performed, the responsible parties, the schedule, and many other items must be prepared. For circumstances where a design firm prepares the scope, the firm must develop standardized forms and systems for this activity. The actual scope of services may be developed by any one of several sources, including:
   a. The owner/client or facilities manager
   b. The design firm
   c. A third party (such as a construction manager, program manager, etc.)

   A scope of services provides a contractual listing of the service obligations of the design firm. Items that are unclear or open for interpretation may create circumstances where the engineer or architect is not adequately compensated for their work, may delay the project, could result in disputes or litigation, or lead to other unforeseen consequences. Scope of services issues will be discussed in more detail later in this chapter.
3. **Project design budgets**: It is vital for architects and engineers to completely prepare a project design budget. Owners/clients will require designers to justify their fee and to support the cost of a fee increase for scope of service changes. Budgeting also provides a guideline for monitoring and controlling the hours and dollars spent in completing the project work. In addition, the project budgeting process:
   a. Allows the PM to obtain a *buy-in* from technical/design staff regarding their scope/schedule/budget items
   b. Requires the PM to think through who will work on the project and when
   c. Allows the PM to closely examine the proposed scope of services
   d. Requires forward pricing fee estimates on long-term contracts to compensate for overhead increases, staff raises, inflation, etc.

   The PM leads the effort to prepare a project budget. Typically, they turn to the technical staff to prepare specific portions. It is the PM who requires the technical staff to justify their own budgets, finalizes the budget, presents it to the client, and negotiates any necessary changes. A poorly prepared budget may result in a substantial financial loss.

Summary

Unfortunately, in the rush or excitement to begin a new project, many engineers and architects fail to adequately evaluate the challenge ahead and objectively examine the cost/benefit of the work. Failure to evaluate your own capabilities may damage your reputation, hurt your bottom line, lead to poor service, and lead to low morale among your PMs and staff.

CONTRACT TYPES

Types of Design Services Contracts

The selection of the proper contract type can be of critical importance to the success of a project. Often, the contract negotiation phase is the most important stage in determining whether a project will be profitable. Design professionals generally have a particular contract type they prefer and they should try to negotiate this type with their clients wherever applicable. The following is a brief description of some of the most common contract types and the advantages and disadvantages of each.

1. **Time-and-materials (hourly charges plus expenses):** This is a common type of contract where the scope of the work is not well-defined. The design professional works at a rate that includes direct salary cost, payroll burden expense (employee taxes and insurance), general and administrative overhead (all other indirect costs), and profit to arrive at an hourly billing rate times the number of hours worked. Project expenses are billed separately—sometimes at a markup.

    This type of contract arrangement guarantees a profit to the extent that design professionals can charge for all their hours and the hourly rate has factored in a profit figure. The disadvantage of this type of contract is that few clients are willing to give a design professional a blank check. As a result, these contracts are usually written to include a stated maximum amount. It is important for both parties to understand whether this maximum is an estimated amount or a figure that cannot be exceeded without prior approval by the client.

2. **Lump-sum:** Contracts that provide for the design professional to perform a certain scope of work for a lump sum are widely used. This type of agreement affords protection to the client and gives

design professionals a guaranteed sum for their work. Regardless of whether the work takes more or less time, the lump sum is paid unless there are changes to the scope of the work. In this event, a new amount is agreed upon. Lump-sum contracts are effective for both parties as long as the design professional is experienced with the type of work required and can estimate costs correctly. It is important to include a contingency factor in the lump sum to protect the firm from the unexpected.

3. **Cost-plus-fixed-fee:** This type of contract is popular for government work and, in theory, should guarantee that costs are recovered and a fee or profit earned. Costs are determined by using a provisional overhead rate while the work is being performed. The actual overhead rate is then determined at the end of the contract or at the close of the fiscal year in the case of contracts that extend beyond one year. Adjustments to the provisional overhead rate are made at that time.

   A problem with cost-plus-fixed-fee work in government contracting is that the government operates on an appropriations basis that ascribes certain costs to different projects. If the appropriation is used up, no further funds can be allocated to a project without considerable justification and paperwork, even if the reason for the extra funds can be explained and documented. As a result, the appropriation generally becomes the spending limit on that project. Therefore, if design professionals spend less than their estimated costs, they only receive reimbursement for their actual costs. However, if they spend more, the extra costs are not reimbursed and must be covered out of their fixed fee. In theory, design professionals are still entitled to the fixed fee even if their costs are less. However, in practice, this may raise the question of whether a scope change limited their involvement and therefore the full fixed fee may be in jeopardy. In addition to cost-plus-fixed-fee, there are variations to this contract type that provide for various incentive fee arrangements.

4. **Multiplier-times-salary:** This method is quite common for design contracts. The multiplier is a rate that covers payroll-burden expense, general and administrative overhead, and a profit factor. It is figured by multiplying the individual's hourly salary rate times the hours worked to arrive at a billing amount for labor. Direct project expenses are billed at cost or at a markup.

The multiplier method is easy to use and assures that all costs are covered. The difference between the multiplier and the hourly rate used in time-and-materials contracts is that the latter often groups classes of employees together and bills them at an average hourly rate. A possible advantage of using average rates over a multiplier is that individual salaries are not disclosed. A disadvantage is that average rates quickly get out of date unless they are revised frequently as employees receive salary increases. This can create a situation where substantial losses occur on a project because of the variance between actual and planned hourly rates.

5. **Percentage-of-construction-cost:** Contracts based on a percentage of construction costs are still used by some firms. Their popularity is fading as both clients and design professionals recognize that these contracts bear no relationship to the cost of work or the amount of creativity required in a project. Nevertheless, many firms are able to achieve satisfactory results on percentage-of-construction-cost contracts and willingly accept this arrangement.

6. **Value-of-service:** One type of contract that is difficult to price is based on the value of the service to the client. For example, engineers who save their clients considerable money through the design of an energy-efficient building are obviously worth more than the value of their time. Design professionals should consider the value of services in developing their pricing structure to the extent that this is possible in a competitive environment.

Summary

There are many variations on these standard contract types and frequently more than one type will be used. For example, until the scope of work can be clearly defined, a firm may work on a time and materials basis. Then the contract will be converted to a lump sum or other basis for the remainder of the work.

BILLING AND COLLECTION

General Comments

Many design firms hinder the collection of their accounts receivable amounts even before signing the contract. They fail to ask owners/clients fundamental questions concerning how they are to bill for services and how they are to be paid. In most design firms, it is the principal or PM who should be responsible for improving the accounts receivable collection process. Unfortunately, they often do not consider regular billing and collection to be an important issue or part of their job description. As a result, many design firms pay insufficient attention to the billing and collection process prior to or just after the contract signing. Basic questions must be answered that will improve the collection process:

1. **What is the client's payment cycle?** If, for example, a client regularly processes invoices on the 25th and the design firm doesn't bill until the 30th, the invoice will be delayed at least three weeks until the client's next payment period.
2. **Does the client require a special billing form?** Failure to use the proper form will delay payment. In many situations, the design firm may not be informed of the reason for delayed payment for several months, thus allowing several invoices to back up.
3. **Is an audit required on each invoice?** Some owners/clients, on larger projects, will require an in-house audit of each invoice. Failure to anticipate and prepare for an audit will often delay processing of the invoice.
4. **Does the contract call for inclusion of supporting material (such as copies of time sheets or vendor bills) along with each invoice?** Failure to enclose this backup will not only delay the invoice, but will create additional work for the firm's support staff who must reassemble this material from files. Anticipating this need would allow for assembly of a billing copy during normal processing of these items.
5. **Who should receive the invoice?** Many owners/clients, particularly on larger projects, separate the invoice approval and processing function from the normal project administration functions of their staff. Sending the invoice to the client's PM, if he or she is not the approval authority, may significantly delay payment.
6. **Before signing the contract, negotiate interest penalties for delayed payment**. Many clients will agree to this after 30 days, and in the case of the U.S. Government, federal agencies must pay interest on past due accounts (over 30 days) if certain billing conditions are met.
7. **Inform clients whom they are to contact regarding questions on design invoices.** It is usually best to have all questions addressed to the individual in charge of the project (typically the PM).

There are many other techniques to improve the collection part of the process. For example, where appropriate, a note from the PM along with the invoice may help create the feeling of a personal obligation on the part of the client. This technique, however, will have little effect where the project administrative and approval individuals are different. Regular follow-up on invoices is vital. The first contact should be made within a week to 10 days after mailing (or e-mailing) to assure its arrival and to respond to any questions. Regularly prepare and distribute accounts receivable aging reports to keep PMs informed as to the status of billings on projects.

Regular billing allows the client to perceive the design firm as a businesslike organization that expects to be paid on time. There is, however, no substitute for proper front-end planning and follow-through. Figure 9.1 provides a suggested design-services billing checklist form.

In situations where the design firm is unable to promptly collect accounts receivable, there are actions that must be taken. Many firms do not negotiate interest penalties for slow-paying clients. For most well-run businesses, including an interest clause would seem an obvious precaution. Unfortunately, many design firm managers believe that since interest may not be collectible, it's not worth charging. This argument is weak, in that interest charges should be used as encouragement for prompt payment and thus become a device to stimulate client action. In addition, in litigation, where an interest rate is not stipulated in a contract, a court may dictate an interest rate. With fluctuating interest rates, the court-ordered rate may not be satisfactory to the design firm. In a Birnberg & Associates informal survey of design firms, less than one-third of firms regularly charged interest on delinquent accounts receivable. The most commonly charged amount was 1.5 percent per month, with the range being from 1 to 2 percent per month. Firms most commonly began charging interest after 30 days.

PROMPT PAYMENT OF CONSULTANTS

Issues

One of the most divisive issues between architects as prime consultants and their engineering sub-consultants concerns payment for services rendered by engineers for architects. The prime consultant/sub-consultant relationship requires trust and businesslike behavior on the part of all parties. Unfortunately, prime consultants are not always prompt in

Contract Management/Project Administration   159

Date prepared: _____
Project no.: _____   Or change order no.: _____
Project name: _____
Client name and address: _____
Client contact for billing questions: _____
Professionals contact for billing questions: _____
Send invoices to (include name and address): _____
_____

Others to receive copies of invoices (include name and address): _____
_____
_____

General questions:
  1. Client billing form required?   Yes (if yes, attached copy)   No

  2. Backup required
       —All vendor and consultant invoices?   Yes   No
       —Vendors only?   Yes   No
       —Consultants only?   Yes   No
       —Other: Yes   No

  3. Timesheet copies required?   Yes   No

  4. Audit required?   Yes   No

  5. Normal client invoice processing date(s): _____

  6. Other: _____

Specific questions:
  1. Fee basis:
       —Multiple of direct salary expense _____
       —Multiple of direct personal expense _____
       —Professional fee plus expenses _____
       —Percentage of construction cost _____
       —Percentage of construction cost _____
       —Fixed amount _____
       —Hourly billing rates _____
       —Other (explain): _____

  2. Maximum fee: $ _____

  3. Reimbursable maximum (if any): $ _____

  4. Errors and omissions project insurance amount to be invoiced: $ _____

  5. Reimbursable markup percentages: _____ %
       All items equal?   Yes   No
       If different percentages to be used, list them _____
       _____

  6. Interest on delinquent receivables: percentage per month: _____ %
       After how many days from the invoice date?: _____ days

Client: _____   Design firm: _____
Approved: _____   Approved: _____
Date: _____   Date: _____

Figure 9.1   Design services billing checklist

paying sub-consultants and are occasionally dishonest in their reasons for slow or nonpayment.

Architects and engineers normally run about 60- to 70-day-average collection periods. The typical design firm has more than one-half of its total assets tied up in accounts receivable and only about 2 to 3 percent of assets in cash. As a result, operations can be impaired and the ability to pay accounts payable limited. Non- or slow payment of obligations to sub-consultants allows the prime consultant to shift limited resources toward immediate needs such as payroll, rent, and utilities.

In recent years, many engineers have moved away from serving as sub-consultants to architects and contract directly with the client, program managers, construction managers, facilities managers, and other owner representatives. This shift has occurred for many reasons, not the least of which is the impact on the professional liability insurance premiums of prime consultants. Premiums are based on total revenues including consultant fees run through the architect's books. By having engineers contract directly with clients or their representatives, the architect's total revenues more closely track his or her own work and are not inflated with pass-through items such as consultant fees. This reduces premiums and can potentially reduce the risk associated with errors and omissions claims to the architect.

Another reason engineers have sought to contract directly with the client is related to the problem of obtaining prompt payment from the prime consultant architect. This is a serious enough issue for several states to address the problem through legislation. In one northeastern U.S. state, the engineering professional societies went to the legislature for relief. The proposed law would have required architects to pay engineers within five days of receipt of part or full payment from a client. The aim was to establish a legal relationship similar to that of the general contractor/subcontractor found in construction firms. Opposition to the legislation was led by the state architectural association who was successful in defeating the proposal. Their primary argument was that the relationship between architects and engineers is based upon that of professionals, licensed by the state and governed by an existing set of rules, regulations, and laws. The proposed legislation was deemed unnecessary.

In Arizona, H.B. 2502 was passed with the support of the Arizona Society of Professional Engineers and the Arizona Consulting Engineers Council. The bill required professionals to pay their sub-consultants within a week of receiving payment from their clients. It also authorized the State Board of Technical Registration to discipline licensed professionals who failed to pay their collaborating registered professionals within seven calendar-days of receiving payment. An exception was permitted where the contract specified otherwise. A *collaborating registered professional* was defined as a firm with whom the prime consultant had a contract to perform professional services and applied to engineers, architects, surveyors, landscape architects, and several other professions. The goal of the legislation was to expedite payment by prime consultants to sub-consultants.

How the Owner/Client Is Impacted

The concerns of engineers are obvious. If they perform work and invoice the architect in accordance with the contract, they expect to be promptly paid. Most contracts contain language that stipulates that the engineer be paid when and if the architect is paid. As noted, some architects choose to ignore this requirement in full or part. Why should the owner/client care about what may be the architect's and engineer's business?

An engineer who is not paid promptly may slow his or her work on a project until invoices are current. Some may threaten or actually stop work. Performance may suffer as staff members are shifted toward projects where payment for services is more promptly forthcoming. A delayed project can be devastating for a client who may be paying substantial financing costs or may incur significant opportunity costs for an incomplete project.

What Can Owners/Clients Do?

There are solutions to the problem of slow payment of subcontractors by prime consultants. The most obvious is the one taken by many engineers—contract directly with the client. Payment is then issued directly by the client and made payable to the engineer. In other cases where the prime consultant/subcontractor relationship is traditional, the client may still wish to issue checks directly to the sub-consultant. Contracts between the client and the architect can be written to include a clause requiring prompt payment to sub-consultants. At the very least, the client should make it clear to both the architect and the engineers that they expect matters such as this to be handled in a proper and businesslike manner.

CASE STUDY: LOOP ARCHITECTS

In the 1990s, two principals of a large Chicago-based architectural firm held a minority interest in two engineering consultants doing a substantial part of the firm's mechanical, electrical, and structural design work. The architectural firm was not particularly profitable and had continual cash flow problems brought on by high overhead levels and some poor business decisions. Most contracts were written in the traditional manner where the architectural firm was the prime consultant and the engineering firms were sub-consultants. Invoices were regularly sent to clients for the work of the architect and engineers. Most clients paid reasonably promptly.

The architectural firm would use the gross payment received to meet payroll and other obligations, leaving little or no money to pay the sub-consultants. This was a habitual practice until a substantial accounts-payable balance was due to the sub-consultants. The two principals rationalized this practice by noting their ownership interest in the engineering firms. They felt entitled to delay or avoid payment on the grounds that they were simply not paying themselves. This obviously neglected the interests of the other owners of the engineering firms. Eventually an agreement was achieved to pay only a portion of the obligation to the engineers. Not long afterwards the minority partners (the architects) were bought out and the relationship between the firms terminated.

SCOPE MANAGEMENT

Key Concepts

Ineffective scope management by design firm PMs is a major factor penalizing project profitability Scope management is the process by which a PM has an in-depth knowledge of the contractual terms of a project and effectively monitors activities that deliver the promised services. Some engineers and architects complain that their clients *nickel and dime* them into insolvency. This occurs when seemingly minor client requests accumulate or expand until they significantly erode a designer's profits. In other cases, the designer is to blame. A poor job of evaluating the needs of the project during the negotiating stage can later necessitate extensive unanticipated work. Scope management is the process by which these activities and requests are controlled. No well-managed client or design firm can afford not to have an effective scope management program in place. (For additional information on scope management, see Chapter 6 and Figure 6.3.) There are numerous steps required to implement this program:

1. **Do an effective job of pre-planning on every project:** This requires detailed checklists of every project activity. Every design firm must have a complete and accessible database of cost and time records on past projects. This information is invaluable when analyzing a new project for potential pitfalls in the budgeting process and for comparisons with current budget and scope expectations.
2. **A system must be in place to monitor costs and time on every project:** It is unacceptable for even the smallest design firm to not have such a system in the office. Many computer-based systems are available. Manual systems are simply too unwieldy, too slow, and often too improperly designed to provide the needed information. A job cost reporting system must allow for coding and monitoring of changes to the basic scope of services. This is often handled by providing for a separate project number under the base job number. All time and costs relative to the change are then charged to this separate number. These changes are not always generated by client request and may be caused by circumstances in the design office.
3. **The individual who is in charge of the project within the design firm must know the scope in detail:** In most offices, this person is the PM. In small design firms, however, the principal or partners may assume this role. Unfortunately, these people may be stretched thin and may not properly monitor staff activities to ensure that they fall within the contractual scope.
4. **Changes in the scope of services must be monitored:** As mentioned before, an effective monitoring system must exist and be made available on a timely basis. It is an excellent policy to establish a separate record of time and costs if there is any doubt whether an activity falls under the basic scope of services. Consolidating segregated information with the base project data is far easier than attempting to go back and separate it from the base job. Some firms even prepare budgets for these segregated change orders, although in many cases the full extent of the work required may not be known in advance.
5. **Time sheet management is an essential activity if scope management is to be successful:** With a proliferation of change order

numbers to a base project, staff may become confused as to where to charge their time. A simple process to control this requires several steps:
   a. A summary sheet listing all projects and change orders open to time charges should be available online for all staff member's reference at the beginning of each time sheet period.
   b. Every PM must be required to review the time sheets of each employee working on his or her jobs before it is accepted. This may be impractical in larger firms.
   c. The computer system must provide PMs with a detailed list of time charges for each of their jobs and change orders. They must examine this information carefully.
   d. Time sheets must be collected daily to improve accuracy. Electronic time sheets must be used to expedite this process.

   A design firm's profit margin depends on the quality of its time sheet management. Principals are usually the worst offenders in not completing their time sheets. To be effective, senior staff must take the lead.

6. **A communication system must be developed to inform all parties of requested or required changes to the basic scope of services:** No matter the source of a scope change, all affected individuals must be notified on a timely basis. In many firms, this is done either verbally (e.g., telephone and meetings) or by lengthy letters or memos, mailed or e-mailed. These are often ineffective and leave a great deal to possible misunderstanding. To be effective, this process must be documented, but in a quick and easy manner. Many firms develop a series of shorthand work authorization forms that fill this need and can be sent to consultants, clients, contractors, and so on. Frequently, these forms become part of the contract between the owner and designer and may require the client's signature before any work is undertaken. This process will be discussed later in this chapter.

7. **Develop the ability to invoice separately for change orders:** In many design firms, small changes to the project scope are absorbed as a marketing cost. However, if care is not exercised, seemingly small changes may grow significantly and reduce profits. Many of these may need to be reviewed with owners/clients on a project by project basis. This issue will be discussed more later in this chapter.

Design Contract Change Orders

The process of managing project change should be a relatively simple one. Unfortunately, many firms fail to adequately prepare a system to cope with these change orders (changes to base project), extras (additions to basic scope), or out-of-scope items (those whose status is not yet resolved). This failure may result in lost opportunities for additional compensation, client conflicts, and problems with outside consultants.

Several simple procedures must be established to manage the change process. Many firms without adequate project control systems do not segregate time and expenses incurred on projects with extra or changed items. Well-managed design firms have built into their job cost accounting systems the capability of segregating changes authorized by their clients or items that may require future discussion. Typically, a suffix is added to the base project number to designate these items. Where the client authorizes the change, and a fee is agreed to, a budget, separate from the base project should be established. A regular project monitoring process must be instituted so that the PM receives a timely status report on project changes or extras. In a situation where the manager believes an item to be beyond the basic project scope, a separate job cost accounting record must be established until the fee and scope status of this work is resolved with the client.

A general rule used by many firms is that if there is any doubt concerning the scope status of the work, a separate job cost accounting record is automatically established. In many cases, a seemingly minor change that might normally be provided as a courtesy to the client grows into a significant cost to the design firm. By immediately segregating these costs, the information is available if the firm wishes to seek additional compensation from the client. Clearly, it is much easier to segregate these costs initially than it is to attempt to pull them out of consolidated records later.

To be successful in managing project change orders, there must be an individual in charge who thoroughly understands the basic scope of services. Where no single individual is in full charge of the project, many such items may simply be lumped into the basic project cost records. Obviously, this directly impacts the base project budget and profit potential. If design consultants do not have a well-established change order management process, then the client or facility manager must impose their own system on consultants.

Communicating Design Change Orders

The most important aspect of successfully managing project change orders is effective communication. Clients, consultants, and staff must all be kept informed of the status of any items that are either authorized or open to future discussion and resolution.

1. **Owners/Clients:** In many situations, designers and clients disagree over whether a change item or extra was authorized and how much the firm was to be paid for its work. Clients may claim that they were *only thinking out loud* and never actually authorized the designer to do the work. They may also claim that they never agreed to pay any additional compensation to the design firm. In still other situations, the designer may feel that a requested change is too minor at present (it may grow!) to request additional compensation.

    To document these situations and to inform their clients, some firms have developed shorthand forms generically called work authorization (WA) forms (see Figure 9.2). Normally, the designer completes a WA form for any activity beyond the basic scope and sends it to the client as a matter of record or for his or her signature. This provides notification and documentation at an early stage before extensive time and costs have been incurred. As a result, a detailed scope for the change can be established, fees determined (where necessary), method of payment outlined, and schedules set. Some clients, such as many government agencies, have their own versions of these forms and no design work can be authorized until all parties sign off on this supplementary agreement.

2. **Consultants:** Coordinating extras and change orders with consultants to the prime design firm can be a difficult task. Since these items are often not covered under the basic agreement with the consultant, questions as to scope, method of payment, amount of payment, and schedule will arise. Situations occur where the consultant proceeds with work possibly beyond that required by the client or prime design firm. On occasion, the amount billed to the prime design firm by the consultant exceeds that paid by the client. One common solution is the use of a consultant WA form similar in nature to that used with clients (see Figure 9.3). This form is the consultant's agreement regarding the work to be performed and covers fees, schedule, method of payment, and so on.

Work Authorization

Client:	Base job number:_____
Project description:	Charge time to:_____
Location of project:	Date issued:_____
Project title:	Initiated by: ☐ Client ☐ Firm
Description of work to be performed:	Reference:
	Starting date:_____
	Est. Compl. Date:_____
	Phase of work: ☐ 0 Design; Prelim & planning ☐ 1 Working drawings ☐ 2 Const. & shop drawings ☐ __Work order number ☐ __Special _____
Remarks	Distribution (as checked): ☐ Client contact: ☐ Design ☐ Architectural ☐ Job captain ☐ Cost consultant ☐ Specifications ☐ Structural ☐ Mechanical
Billing instructions: ☐ Included in basic fee ☐ Change and/or additional service ☐ Not determined ☐ Special ☐ Maximum compensation, if any $_____ .	☐ Field superintendent ☐ Consultant ☒ Managing principal ☒ Comptroller ☒ Accounting ☒ Main file ☐ Other_____
Project manager to provide complete information and check applicable boxes, including appropriate distribution of copies.	☒ Project manager: _____ Authorized
This project work authorization is subject to all terms and conditions of the contract signed (date): _____.	☒ Principal in charge: _____
Client approval:_____ Date:_____	Approved

Figure 9.2 Work authorization form

Consultant Work Authorization Form

Consultant:

Client:

Project title: Job no. _____

Estimated completion date: _____

Instructions: This agreement is subject to and governed by all the terms and conditions of our agreement dated _____ entered into by the undersigned unless modified In writing.

Scope of work:

Special provisions:

Fee (including the terms of payment):

Date:_____ Date:_____

Prime design firm Consultant

Figure 9.3 Consultant work authorization form

3. **Staff:** For projects with numerous change orders or extras, problems arise in keeping the staff informed of proper time and expense charges. Without accurate input, the project job cost records will be much less useful. To reduce this problem, some firms distribute paper or electronic copies of the WA forms or a summary of all active change orders to key staff members.

It is the responsibility of PMs to monitor the time charges of staff working for them. Effective management of project change orders requires a system and the discipline to follow the procedures mandated by the system. It is important to keep separate records of all extras, change orders, and out-of-scope items. Without these records, the design firm will lose opportunities to obtain compensation for all actual work done and will likely have lower profits.

CONTROLLING PROJECT DESIGN COSTS AND SCHEDULES

The Challenge

Controlling project design costs is among the most challenging of all PM responsibilities. This material provides a review of information from other chapters highlighting approaches and tools to keep project design costs within budget. It discusses various methods to organize a design firm for project delivery, reviews some important items to keep in mind before signing a design services contract, evaluates causes of projects running over budget, and reviews various cost control tools.

1. **Design firm organization:** How a design firm organizes internally for project delivery significantly affects its ability to control design costs. The individual managing the project is the major point of contact for clients; controls the project scope, schedule, and design budget; and is the only individual involved in the project from beginning to end. If there is no PM, or if a senior manager is attempting to manage projects and the firm, then client service and satisfaction are in doubt. To recap material discussed earlier in this book:
    a. Matrix/strong project management system
        i. Provides one individual to monitor and manage the project

ii. One manager from project start to finish
iii. Continuous point of contact for client
iv. Requires equality of authority and responsibility to be effective
v. Provides one individual to control project design costs
vi. Encourages decision making at lowest effective level
vii. Flexible system that allows for change
b. Pyramidal/departmental/other systems
i. No single individual monitoring and managing project costs
ii. Pyramidal usually leads to crisis management of projects
iii. Departmental fails to provide one point of contact for clients

Pyramidal organizations are headed by a principal in a design firm, or an executive or bureaucrat in a corporation, institution, or government agency. This individual has a variety of responsibilities and is unable to devote full-time to managing projects. Crisis management is usually the result when a pyramidal structure exists for managing jobs. The ability to meet design and construction budgets, schedules, and other project requirements is difficult.

Departmental organizations lack one project leader and instead pass project management responsibility from department to department and manager to manager. Each department establishes its own priorities, schedule, and budget. Communication is made more difficult by the involvement of several department heads.

In multidisciplinary design firms, a matrix system is typically used within the discipline/department where a PM remains in charge of the entire project no matter what discipline the project is in. Matrix organizations provide for one project leader with equal levels of responsibility and authority for the project scope of services, design budget (if a design firm), and the project schedule. The PM is the point of continuity for the entire project team and is the central communication link for the project in your office. Equality of authority and responsibility allows the PM to make decisions quickly and effectively. This allows projects to be completed with fewer delays and communication problems resulting in design cost savings.

2. **Effective communications:** The PM is the key communications link in a design organization. He or she must provide and

receive information from all other members of the project team. The PM serves as the primary point of contact in his or her respective organization. If the PM position does not exist, or lacks the necessary tools, the flow of information is restricted or blocked.

The concept of Partnering (discussed in detail later in this chapter) has existed for a number of years. It requires early meetings to set the stage for a successful relationship between the team members. This early opening of the lines of communication is continued throughout the project and greatly improves the team's ability to meet scope, schedule, and budget requirements. The PM is the key individual in the Partnering process by his or her regular participation in Partnering meetings.

Scope creep is a particularly challenging issue. PMs are reluctant to say no to their clients and do not want to risk damaging a relationship by charging for every item of service. Unfortunately, the cumulative impact of these items can be significant and can reduce a project's profit potential. A balance must be achieved by informing a client when an item is considered beyond the contractual scope of services while judiciously charging for appropriate items.

3. **Monitoring the project's progress:** Many tools exist for monitoring the progress of a design project. The most important is the actual cost versus budget report. This is a fundamental tool for every PM and should be available on demand. It is essential for a design firm to have an effective job cost reporting system to keep PMs up-to-date on their project's status. Of course, no system will help if the PMs do not understand the project status reports and, therefore, do not take timely action when necessary. Additionally, managers must understand how to prepare a project budget, how to compare actual costs versus budgets, when to set up separate records for out-of-scope items, and a myriad of other actions.

An effective monitoring system allows the PM to analyze variations from the budget plan and take necessary corrective action. This may require adding staff to complete the project on time, changing staff to those better able to complete the work, revising budgets to reflect the reality of the present situation, evaluating the project for scope creep, or revising schedules.

WHY PROJECTS RUN OVER THEIR DESIGN BUDGET

General Comments

Design firms are in business to provide solutions to their client's needs and to make a profit. If the PM is not successful in controlling project design costs, then he or she will have likely failed to achieve the latter goal. Often, client service suffers if a project is running over the planned budget, and the result may be a failure to provide a satisfactory solution for your client's needs. Some things that may interfere with the success of the PM are:

1. **Meddling by senior managers:** This often results in the diminishing of the PM in the eyes of the project team. Control of scope, schedule, and budget may be lessened or removed from the PM as a result of meddling. When this occurs, the project management system is compromised and will result in poor communications, multiple points of contact for clients and consultants, lack of decision making by the (nominal) PM, and often crisis management by principals.
2. **Excess perfection:** This can afflict designers who delay projects by not committing to a design solution, technical staff who burn up the design firm's budget by over detailing project drawings, specifiers who select products beyond the budgetary and performance requirements of clients, clients who have unrealistic expectations of building performance, etc. PMs must learn techniques to control excess perfection wherever possible. Common techniques include making certain the staff knows project expectations, understands the scope of services required, and has participated in the preparation of the schedule and design fee budget.
3. **Overburdened project managers:** When PMs are overburdened, they can become crisis managers who are unable to fully study issues and problems, who lack the time to properly monitor a project's status, and who fail to communicate effectively with all team members. This crisis management situation can quickly harm service and the ability to meet project design budgets. One major reason that a PM might become overburdened is due to a large number of design change orders/scope changes that can turn one project into many. Cross-training of staff may provide

capable assistant PMs or other individuals to relieve some of the burden.

4. **Working outside of the agreed upon scope of services:** Change orders are common, but it is important to avoid working outside of the agreed upon scope of services. In an effort to please clients, some PMs agree to do tasks without adequate/any compensation. Unfortunately, seemingly small changes may grow into costly items for the design firm. Tools must be developed to manage scope changes (see Figure 9.2 in this chapter). If needed changes are not promptly communicated to other team members they may continue working based upon previous directives. This may result in the need to redo work and wastes time and the project design budget. The PM is responsible for communicating changes to other team members. In particular, the need for information from the client must be anticipated. Equally important is anticipating the need for client decisions.

5. **Poor project documentation:** Poor documentation makes it difficult for the PM and other team members to quickly, accurately, and completely communicate information to each other. Errors, omissions, and poor drawing coordination may result. Historical records may be incomplete or inaccurate causing future problems long after the project is complete. In the short term, rework may be necessary because a team member did not receive information regarding changes, decisions, or errors.

Turning Problem Projects Around

Perhaps the most difficult challenge for PMs is the identification of problems. This is not always easy when the manager is close to the project. Others may need to help in this process—particularly the client—as they may have insight based on their previous projects.

There are several other steps to turning around a troubled project. Once the source of the difficulty has been determined, the PM needs to set priorities—decide what steps need to be undertaken first. It is important to examine the scope to determine if the work is beyond what was agreed under the contract. A job cost reporting system is crucial to the early discovery of problems. Once issues are determined, it may become necessary to adjust the staff assigned to the project. In particular, the PM may be overburdened and other PMs or assistant PMs may need to help. If feasible and required, the technical/design staff may need to be changed. When a project is seriously behind schedule and/or over budget, staff may need to be added if available. It may also become necessary to revise the design budget to reflect the new/current conditions. Revised time schedules should be set and a buy-in obtained from the project staff for the proposed solutions/new budgets and schedules.

PROJECT ADMINISTRATIVE ACTIVITIES*

Filing Project Paper Data

All projects generate huge amounts of paper and electronic files with varying degrees of value. Locating a particular item can be difficult since individuals often develop their own customized filing system. Although any system will work better than no system at all, good systems have certain principles in common. Appoint one person to be in charge of the filing system. Duplicate files and duplicate copies in project files should be avoided. Sort your data into similar categories. Bind data into volumes and do not permit individual items to be separated from the group. Provide a sign-out system so that missing volumes can be located. Check frequently to be sure the file is being kept up-to-date. It is common for filing to be a free-time activity. Unfortunately, far more time is lost looking for improperly filed information than is spent in maintaining a properly planned file on a regular basis.

To make project files usable, they must be broken down into sections of manageable size. Any grouping larger than what can be placed in a single folder is too large for efficient locating of specific information. A file should not be subdivided more than necessary to keep the subdivisions of manageable size. The volume of material accumulated on a previous, similar project is a good guide to the amount that is likely to be generated by a new project. Avoid elaborate cross-indexing systems. Simple chronological sequencing is usually the best way to locate documents quickly since most individuals searching a file usually have a general idea of the time a document was received or created.

Separate incoming from outgoing correspondence. Don't bother trying to file letters that answer questions raised by earlier correspondence with the earlier letter. Many letters are never answered directly or answers to several letters may be included in a single response. Many important

---

*The material in this section was mostly prepared by Gene Montgomery, AIA, with additional material by the author.

e-mails should be printed and filed as well. Every organization must develop policies for handling and recording these important items.

Incoming Correspondence

All incoming correspondence should be routed to the person in charge of the project. Some mail may need to be rerouted to others on the project team. In most cases it is wise to make copies of the document and distribute the copies to persons listed on a standard distribution list. Circulating a single copy to several persons is rarely successful since the copy usually stays on the desk of the first to receive it. Distributing on a need-to-know basis can also be a problem as the person in charge may not recognize the importance of a document to another member of the team. To reduce the need for copying long or bulky documents, a memo or e-mail acknowledging receipt and indicating who has custody of the item can be circulated in lieu of the actual document. Zoning codes or soil boring reports, which may be important to several team members, are often handled in this manner. It is important that the original document be added to the permanent file as soon as possible to reduce the possibility of loss.

The purpose of distributing copies is to keep the project team informed. Since the original can always be consulted, recipients should only save those documents that they will be using regularly. Technical staff members generally have no need to maintain their own job files.

Outgoing Correspondence

Many firms require the PM to sign or review all outgoing correspondence (snail mail and e-mail). Even seemingly routine inquiries to manufacturers may violate office policy or client requests. A copy of all outgoing correspondence (snail mail and e-mail) should be kept in a central project file and a separate chronological correspondence file may also be helpful. Copies should be distributed to team members in the same manner as incoming correspondence. Your attorney or other advisor should be consulted for guidance on retention practices.

Interoffice/Intraoffice Memos

It is often convenient to distribute job information by memo or e-mail. Like any document related to a job, these are subject to discovery (legal) proceedings in the event of lawsuits. As a result, judgment must be used when preparing memos. For example, an internal memo documenting a project decision might be used to find the design firm at fault in a professional liability suit.

Project Meeting Notes

Many firms find it advantageous to prepare a memo to record the minutes of job meetings. Most projects involve many meetings with the owner, consultants, contractors, government agencies, and others. Often, decisions reached are not clearly understood by all parties. By circulating minutes to all participants, the writer, in a sense, controls the decision-making process. Months later these minutes are invaluable for determining the understanding of the parties at the time of the meeting. (For more on meetings, see Chapter 5.)

Work Authorizations

During the progress of a project, members of the team discuss many alternatives, scenarios, budgets, and schedules. Because most people are eager to get on with the job, they often misunderstand the significance of these discussions and proceed to do work based on preliminary assumptions. Frequently, clients do not follow through with written authorizations of work discussed. The team should expect and demand receipt of formal authorization before proceeding with work. (See samples earlier in this chapter.)

Telephone Calls

The telephone is indispensable for communicating information and assisting in problem solving. It has the disadvantage of not providing documentation of decisions and discussions. It is important to record decisions reached by telephone immediately after completing the call. This process can be invaluable if questions on a job arise at a later date.

Manufacturer's Assistance

Many complicated products or systems can only be described and specified with close collaboration between the engineering or architectural designer and a manufacturer. Designers should be careful in these situations since close collaboration can lead to an eroding of the designer-client relationship. Particularly in publicly bid projects, the designer must carefully explain to the sales representative that the manufacturer's assistance cannot imply endorsement of his or her product at the expense of others. Even when carefully explained, bad feelings can result when a designer relies heavily on a manufacturer to prepare designs or specifications. Whenever possible, this practice should be avoided. Designers should employ a knowledgeable consultant or pay the manufacturer to design the system to avoid a potential conflict of interest.

Confidentiality

Designers work for their clients and must keep the best interests of the client in mind at all times. It is important that the PM have an understanding with the client on what information can be released to the public, to the construction industry, or used in the design firm's own marketing materials. Some owners take the somewhat unrealistic position that all information about the job is confidential and cannot be discussed with anyone not directly involved with the project. Although they may not want the designer to release information to the media about the project, most clients recognize that firms must gather information from contractors and salespeople and provide them with data about the project. The best policy is to insist on staff confidentiality, but authorize the person in charge to use discretion on the release of information.

Dealing with Consultants

Most projects require consultants to perform certain tasks. The project staff must understand the duties and responsibility limits of the various consultants. Work authorizations must explain what is expected of a consultant. Schedules should realistically consider when the consultant can begin his or her work and when it must be completed. In addition, effective communication is essential. Job memos and correspondence should be routed to consultants to keep them informed on the progress of the project.

Checklists

A list of items to check before completion of a project can supplement memory and provide a useful record of matters considered. Checklists take many forms and can be used with varying degrees of formality. Many design firms and owner organizations have developed standard preprinted lists of activities to be performed on projects. Unfortunately, checking such a list can become a perfunctory task and may provide a false sense of security by giving the impression that items and activities not listed are unimportant. A comprehensive list by necessity includes so many inapplicable items that all items may become diminished in importance.

Other standard documents can serve as useful checklists. A master specification table of contents, such as the one found in Masterformat published by the Construction Specifications Institute, becomes a type of checklist—reminding the specifier of items that may be required on the project. Probably the most useful checklists are the informal ones prepared by the user. By jotting down a list of items to check as the project proceeds, the user has a handy reminder of specific items that need to be resolved before the job is completed. Job memoranda can also be used as a checklist. Before a project is completed, review old memos and check off all items to be certain that commitments have been kept.

Check Prints

As project teams grow in the number of people involved, it is imperative that each member knows what the others are doing. Up-to-date *check sets* provide this communication. But, check sets can be expensive—not only in the cost of the set, but in the time lost while drawings are plotted. There are ways to decrease the cost of check sets and increase their effectiveness—provide fewer but better check sets. Bind sets with spring fasteners and replace only the sheets that have changed since the last printing. Have several members of the team share sets. Encourage users to note incorrect or out-of-date information on the set. When sheets are exchanged for newer ones, the team leader can review the notes and evaluate their significance. Saving old, marked sheets can provide an audit trail when errors are encountered.

E-mail/Text Messages

E-mail can be extremely helpful in transmitting information. It allows the recipient to read the material at his or her leisure. Unfortunately, the ease of e-mail distribution means that many individuals who do not need the information are flooded with extra reading. Firms need to develop policies regarding distribution of information via email and the printing of e-mails for archiving. Text messages are useful for brief communication, but lack any permanence or proper documentation. Great care should be exercised in the use of text messaging for project communications. Most risk managers would strongly discourage the use of text messaging on projects.

PROJECT NOTEBOOKS AND PROJECT MANAGEMENT MANUALS

Project Notebooks

One very successful organizational technique for PMs is the development of a project notebook for each major job. These binders (or electronic files) contain project information for ready reference, documentation, and portability. Their primary purposes include the following:

1. **Organization of information:** Project notebooks provide a ready format to organize information on projects. Typically, these notebooks contain standard sections (see Figure 9.4) as a framework to follow for every project.
2. **Reference by others:** One major benefit of project notebooks is their ready availability and easy reference by other members of the project team. This helps the PM to achieve one of his or her primary jobs—communicating information to others.
3. **Developing historical databases:** In many engineering and architectural offices, the most complete information is contained in the PM's personal paper or electronic files. By creating project notebooks, information is organized for future reference by others including the current PM, succeeding PMs on related projects,

---

Section One: List of key project contacts (client, consultants, suppliers, contractors, sub-contractors, etc.) including name, firm, address, telephone and fax numbers, e-mail address, secretary/administrative assistant name, home telephone numbers, etc. Also include information such as whom to copy on correspondence, etc.

Section Two: Project scope issues, contracts, copies of change order forms, etc.

Section Three: Budget information such as design fee budgets, supporting documentation, etc.

Section Four: Construction cost data, material information, etc.

Section Five: Scheduling information and charts such as personnel plans, Gantt, CPM, etc.

Section Six: Meeting notes, field notes, etc.

Section Seven: Additional reference material

Section Eight: Some project managers maintain a chronological correspondence file as reference.

Additional: Lessons learned

Figure 9.4 Project notebook outline

and marketers. This material is vital for budgeting future projects, in the event of disputes or litigation, and for general documentation and reference. Some organizations have eliminated paper binders and only maintain this material electronically.

Project Management Manuals

Development

Among the most important training and management tools for PMs is the project management manual. Unfortunately, very few organizations prepare this valuable document. Its general purpose is to document most of the responsibilities and tools required in performing the project management role in a firm. Paper versions of these manuals should be in a three-ring binder format to facilitate its use and to allow for easy revision. Electronic versions are an alternative and should be available to all staff and project team members. Each section should provide for a *sunset* date at which time it is revised. The manual preparation should be managed by a senior executive, while responsibility for research and writing should be assigned to several PMs. The sections must be assigned like a project, with specified due dates and content. A professional editor/writer should prepare the final draft.

Typically, the manual is used as a reference document by experienced PMs, as a training tool for new or less experienced managers, and as an orientation tool for new staff members at the middle and senior levels of the firm. Often, firms provide copies of all or part of their manual to their clients in an effort to inform them of the firm's procedures and methods of operation. This can clarify critical issues in the early stage of a project (see Figure 9.5 for the various reasons project management manuals are developed).

Content

In general, a project management manual should include the following:

1. **The firm's project management philosophy and approach:** This is a brief section on the firm's concept of project management and organization.
2. **The role of the PM in marketing:** This must include an outline of the firm's marketing plan and structure. In addition, the expected interface of the PM and the marketing staff should be clearly explained.

> Project management manuals are developed for many reasons, including:
> 1. As part of the RFP/RFQ process, a potential client may request a copy of the manual initiating a flurry of last minute activity to assemble this previously unavailable material.
> 2. A firm may be seeking or is required to obtain ISO certification. Preparation of a project management manual is an integral part of this process.
> 3. Preparing a project management manual assists in defining project management in a firm (also very useful in facilities management organizations—see the case study beginning on the next page).
> 4. The preparation of a project management manual helps you seek out the "Best Practices" of others.
> 5. The existence of a project management manual helps to enhance internal and external communications by sharing pertinent parts of your manual with staff, clients, consultants, contractors, and others.
> 6. A project management manual is a resource useful in training new or aspiring PMs and helps to define a career path for these individuals.
> 7. A project management manual helps to standardize the activities of current PMs while allowing freedom for individual style and project needs.

Figure 9.5 Project management manuals

3. **Project front-end planning and contractual activities.** This section clarifies the PM's role in preparing the project program and budget and outlines responsibilities and work assignments. In addition, the role of the manager in the writing and negotiating of contracts must be reviewed. Copies of all required forms should be included for illustration purposes.
4. **Management of the project team:** Responsibilities for meetings, communications, personnel issues, and performance appraisals are outlined in detail in this section. Additionally, consultant and client relation activities (including the handling of out-of-scope items) are discussed and copies of all current forms are included. Detailed project checklists and position descriptions for most project roles are also essential.
5. **Quality management:** Although PMs often do not conduct quality reviews themselves, they are responsible for ensuring that a time and budget allocation exists for this activity. They must also require that quality assurance reviews be conducted as scheduled.
6. **Control of budgets and schedules:** Requirements for review of budgets and project status reports must be outlined and standard forms should be included. Standard operating procedures for the

administration of these items (particularly change orders) are outlined here. A procedure for corrective actions in the event of problems should be included.
7. **Project closeout and follow-up activities:** Checklists and procedures for completing a project are reviewed in this section. As in other parts of the manual, this section should include samples of all required forms. Instructions for project follow-up and continued contact must be outlined.

Although enhancements to the manual can be included (such as a section on risk management), these sections provide the basic information required. The key to a successful project management manual lies in its completeness and easy use. Any system or manual that becomes too complex or cumbersome will not be used.

CASE STUDY: INTERNATIONAL POTATO CORPORATION

Background

The International Potato Corporation (IPC) is an Idaho-based, wholly owned subsidiary of Kansas City-based Giant Food Products Corporation. IPC was founded just before World War I and benefited by America's entry into the war in 1917. Large War Department contracts provided IPC with a major market for its potatoes. The boom years of the 1920s furthered the growth of the firm. Even during the Depression, business was good as potatoes provided a cheap and nutritious source of food for the struggling American public. Depressed farm prices also allowed the company to purchase potatoes from farmers at very low cost. When World War II began, IPC continued its growth as the U.S. military needed a huge amount of potato products.

After the war, IPC began to experiment with frozen potato products. Progress was slow as the technology was new and untested, but by the late 1950s there was growing success and widespread distribution. The growth of suburbs, supermarkets, and convenience foods made the new product lines successful. In 1961, IPC acquired the Oregon Potato Producers Corporation (OPPC), adding to its existing plants in Boise and Pocatello, Idaho as well as in Spokane, Washington. OPPC had processing plants in Medford and Salem, Oregon. In 1975, IPC acquired its first non-U.S. operation by purchasing the Drumheller Potato Corporation, based in Alberta, Canada.

New technology allowed IPC to expand its product lines, enhance distribution, and add processing plants. A European operation was added in the 1980s and Asia was added in the early 1990s—including a plant in China to feed that nation's growing appetite for Western food products. In 1997, the third generation owners of IPC sought to retire and found a willing buyer in Giant Foods. Always a closely held and managed business, at the time of the sale, inefficiencies in operations were putting stress on the firm's bottom line.

Facilities Management

IPC currently has 22 processing plants located in five countries: 15 in the U.S.; 3 in Canada; 1 in China; 2 in the U.K.; and 1 in Hungary. These plants range in age from 10 to 50 years. Many are in need of significant updating, however, as they run three shifts, seven days per week, there is little opportunity to do more than replace equipment. Every three weeks, each plant completely shuts down for 48 hours for a total cleaning of all food handling equipment and storage facilities. Despite this regular cleaning, IPC has faced recalls, investigations, and legal actions from consumers and regulators in the U.S. and European Union.

Each plant is staffed by a chief engineer who is primarily responsible for maintaining processing operations. Operation of the plant itself is the responsibility of corporate facilities engineering in Kansas City. While the corporate facilities staff regularly visit the various plants, they largely triage the most significant facilities problems and needs. Corporate budgeting is cumbersome and largely focused on production. The facilities staff is largely made up of mechanical engineers (with only a smattering of other engineering disciplines), a couple of professional facilities managers, and one architect. None have any formal project management training. There are few tools available for use by corporate facilities engineering. Little of the plant information is computerized, manuals and procedures are almost nonexistent, each plant and facility manager develops their own documentation and procedures, there is little institutional *memory* other than that of various individuals, and a myriad of other problems exist.

The facilities group is led by a few long-time senior engineering employees. While well-meaning, these individuals have spent most of their working careers with IPC and have little understanding how other companies operate. Training of the facilities staff and plant managers is largely based on *boots-on-the-ground* issues. There is little focus on how a project/facility manager should function, no real understanding of the need for standardized procedures and documentation, no attempt to pass on experiences to younger staff, and hiring is focused on technical skills almost exclusively. Only recently, IPC took its first steps toward developing a project management manual to guide staff. The draft of this manual indicates a significant focus on technical issues.

IPC is profitable, very busy, and dominates the frozen potato market in North America and Europe. While there is recognition that the plants are often inefficient, aging with little maintenance, and will eventually need to be replaced, nothing is currently being done to prepare for the future. As one executive put it, "It's hard to argue with our success."

PARTNERING

Concepts

Few industries have the history of litigation and disputes found in the construction industry. Many of the contract documents produced by the various professional societies have been drafted specifically to pass risk to others or to spread risk among all parties to the project. Conflict is common; trust is not. Some in the industry have sought to change this and restore trust and a spirit of working together to achieve a common goal. This has become embodied in the Partnering concept.

Tom Warne, former director of the Utah Department of Transportation prepared a manual on Partnering for the American Society of Civil Engineers titled *Partnering for Success*. In this manual, he wrote: "Partnering recognized that there are many stakeholders on any given construction project. A stakeholder may be an owner, a prime contractor, a design engineer or architect, a subcontractor, a supplier, a local community or business group, a governmental agency, or any one of a host of others."

A key to the Partnering process is for all of the parties involved in a construction project to come together and examine their goals and objectives to determine common ground. Typically, these goals and objectives include a desire for quality, profitability, safety, on-time completion, and other similar items. The Partnering team then determines how to work together to achieve these.

This is undertaken initially through a Partnering conference where all of the key players including their respective PMs meet for two or more days. Each organization represented presents their goals, objectives, concerns, methods of operation, and so on. Common ground is identified and agreement is reached on dispute and issue resolution. This is embodied in a Partnering charter that all sign and agree to follow. The initial Partnering conference is best led by an outside facilitator who is objective and can direct the process—*it should never be led by the client*. As part of this conference, the parties agree to establish a regular Partnering meeting process throughout the life of the project. In order to make Partnering a success, several additional items should be considered:

- Partnering requires the strong backing of the client. If he or she believes in it and supports it, then it will succeed.
- The client has to be willing to pay for the Partnering process by hiring a facilitator and by paying for the time of the team members for participating.
- The design and construction team members must support Partnering from the top down. If senior management does not provide support, then the process becomes much more difficult.
- PMs must be regular participants in Partnering meetings.

This book has free material available for download from the Web Added Value™ resource center at *www.jrosspub.com*

10

MANAGING PROJECT QUALITY AND RISK MANAGEMENT

QUALITY AND RISK MANAGEMENT CONCEPTS

Introduction

Many in the construction industry perceive risk management to primarily include writing a good contract and purchasing sufficient liability insurance. While these are basic elements, they are only the shell of a much more sophisticated approach to risk management. A true risk management structure encompasses everyone and everything in your organization. Even the definition of risk management can vary based upon your particular role in the building process. For example, to a building owner risk can include the loss of use of a building because of construction delays or problems. Significant costs may be associated with the inability to occupy a building including lost sales due to inadequate capacity, higher manufacturing costs, higher overhead costs, lost rents, and other opportunity costs. Owners have risks associated with a building that has higher than expected design and construction costs or a building with poor or below expectation functionality.

To an architect or engineer, risk is primarily related to the following: legal action by the owner/client, costs of design change orders with little or no additional fee, damage to reputation due to building technical problems such as a sick building, loss of repeat clients due to perceived poor service, and so on. For engineers and architects, contracts are intended to arrive at an understanding with a client before anything goes wrong, and insurance is intended as protection in the event something does go wrong. Between these two extremes are a wide range of ideas, actions, approaches, preventative measures, and concepts intended to take a great deal of the *risk* out of risk management.

Unfortunately, there have been far too many examples of building failures, construction worker injuries and deaths, sick building problems, and the failure of building systems such as mechanical systems, roofing systems, curtain wall, and material failures. In some cases, these problems have led directly to the failure or closing of design firms. In extreme cases, such as in the Kansas City Hyatt Regency disaster in the 1980s, a client was confronted with enormous long-term legal expense and settlement costs and the engineer suffered the loss of his professional license. Even in less devastating cases, all parties may be faced with huge time commitments to deal with alleged errors and omissions, legal fees, and damage to reputations.

A Total Approach

Good risk management requires a total approach to the issue. Design firms must regularly examine their project delivery methods. Those with systems that create a crisis management environment with overburdened senior managers attempting both firm and project management are at great potential risk. Engineers and architects who fail to develop effective training, cross training, and mentoring programs risk major errors by inexperienced and poorly supervised junior staff or snap decisions made by overburdened project managers (PMs) lacking capable technical staff.

Design firms should avoid seeking work for which they have little experience. While they may be technically qualified, they may not be familiar with the unique requirements of the client type (such as federal agencies), they may lack knowledge of unique codes and regulations (such as found in California), or they may lack a local presence. Geographically distant team members may complicate communications. The University of Washington found this particularly challenging a number of years ago for a new computer sciences/engineering building where the architect was in Boston, the general contractor based in London, Ontario, and the client in Seattle. (See the case study of "A Showpiece-turned-sour..." in Chapter 1.)

Many firms lack adequate systems for supplying PMs and other staff with necessary project information. This lack of project status and communication tools and systems encourages *shoot-from-the-hip* decision making based on *gut feelings*, not facts. The use of project extranets can improve the sharing of information, encourage prompt and accurate decision making, and provide for complete documentation and an audit trail on a project. Building Information Modeling (BIM) is the latest attempt to improve coordination and information sharing, and will be discussed in Chapter 13.

In some owner organizations, initial decisions regarding a building's function, budget, schedule, site, aesthetics, and other factors are made by financial staff, end users, and other staff with little or no input from in-house facilities management, technical staff, or outside consultants. The result may be unrealistic expectations for design and construction schedules and budgets, building performance, or other issues. Corporate, institutional, or agency decision making based upon flawed information can put the client's project at risk before the first line is drawn on paper and could lead to future conflict with engineers, architects, and contractors.

In every construction project a team is assembled. This team typically includes engineers, architects, owners, general contractors, specialty subcontractors, vendors, suppliers, regulators, and many other parties—all of whom will impact the level of risk a firm is exposed to. Your success or failure in communicating on an effective, timely, and accurate basis with all team members can be the difference between low and high risk. For example, an architect's failure to communicate a change decided by the client to consulting engineers can result in rework, delayed design completion, higher costs, and potential construction problems or failures. This failure exposes all team members to a higher level of risk.

Effective Communications

A good risk management program emphasizes effective communication. While PMs are typically the conduits for communication, they do not need to write every letter, memo, or e-mail; attend every meeting; or make every telephone call. Good risk management would allow a PM to review most or all letters and memos before they are sent. In situations where other staff members attend meetings without the PM, they must report on the discussions to the PM using any efficient method. The same would be true for telephone calls.

All staff must be trained in risk management issues and be reminded of these issues at every opportunity. For example, some firms produce listings of *red flag* words (see Figure 10.1) to be used with care in written communications. Examples include the use of words such as *full, complete, all,* and many others. These words cannot and should not be completely avoided, but care must be taken in their use. Never promise the impossible such as *providing complete mechanical engineering services.* Unfortunately, many firms regularly use such terms in their marketing material.

Neither owners nor design firms should rely only on their attorney for risk management advice. These individuals are invaluable in drafting contracts, but they rarely have any concept of how to aid you in your total risk management strategy. Consider incorporating alternative dispute resolution (ADR) devices into your contracts. Litigation and arbitration require the participation of lawyers and are often more expensive, time consuming, and divisive than other ADR devices.

administer	exact	or equal
advise	examination	oversee
allowable	expert	perfect
always	extremely	periodic
any	final	possible
appropriate	full	prevent
approve	guarantee	probable
as required	immediately	proper
assure	inevitably	prove
best	inspect	safe
certify/certification	instantly	shall
clean, pure	insure	should
cleanup	investigation	significant
complete	likely	sound
control	maximize	stable
constant or continuous	minimize	study
critical	most	sufficient
define	must	suitable
detailed	necessary	supervise
determine	never	thorough
direct	no	unsafe
eliminate	non	warranty
ensure	none	will
equal	not less than	worst-case
essential	obvious	zero
estimate	optimize	
every	optimum	

Figure 10.1 Red flag words

Incorporating a formal Partnering process (see Chapter 9) into the project can be an effective method to enhance communication, discuss problems and conflicts, and settle disputes quickly and cost-effectively. Unfortunately, some owners use Partnering as a tool to try to eliminate construction change orders and claims. This is misguided, shortsighted, and inappropriate. A properly structured, effective Partnering process can be a successful risk management tool.

Insurance is a necessary risk management tool, but it does not cover everything. For example, the typical professional liability (errors and omissions) policy doesn't cover express warranties. In a design firm, if you promise your client that a specific result will occur, then you've made an express warranty and your insurance will not cover you in the event of a problem or failure. Your insurance company can be of great help to you in developing a risk management program. Some provide seminars to teach aspects of effective risk management and offer their insured a discount on premiums for attending.

Documentation

There are many other aspects of a good risk management program. The preparation of construction documents must be regularly examined to determine where repeat errors occur. This is particularly true with computer-aided design and drafting (CADD) drawings, where mistakes are repeated from one project to the next because they are in your computer files. The level of detail required on drawings should be examined. Is the architect or engineer providing enough to convey necessary information to contractors, subcontractors, code officials, and so on?

Some owners/clients produce their own newsletters or reports of common design and technical errors they have found in their projects. For example, the U.S. Department of Veterans Affairs (VA) produced the *A/E Quality Alert* to help its consultants *avoid errors, omissions, and costly liability*. As they note, "VA hires the best Architectural/Engineering (A/E) firms in the country and yet we experience repeated errors and omissions on every project. These are not new or unusual items, but they consistently get past A/E and VA quality assurance (QA) reviews and result in construction change orders. Please review your project documents and make sure you do not make these mistakes."

Too many firms focus on quality control. This is the process of developing checking procedures. Errors are made, time and money are allocated to attempting to catch errors, and even more cost is involved in trying to fix the errors. Quality management is a much more proactive and preferred approach focusing on developing good communication, training, and procedures to prevent mistakes, not focus on fixing them after the fact. Many in the industry still use the term QA interchangeably with quality management.

Complete and accurate documentation of all activities is crucial. This includes e-mail, voice mail, and telephone conversations. Meeting and field notes must be prepared immediately and incorporate all discussions, decisions, and issues of importance. Field issues requiring the client's attention must be brought to him or her immediately.

Design change orders, either perceived by the design firm or requested by clients must be documented. These typically take the form of specific requests by clients, perceived out-of-scope items by engineers or architects or those yet undetermined. Failure to document design change orders may result in additional unpaid work for a design firm, client dissatisfaction, and poor communication among project team members regarding changes.

Summary

Risk management is a continual effort. You must understand the nature and extent of the risk you and your firm are exposed to. A comprehensive strategy must be developed to address risk and to remediate its impact. Never take for granted that you are protected by contracts and insurance. Do not assume that because you have never been sued, you will not be in the future.

QUALITY MANAGEMENT

Introduction

Few design firms engage in any regular or formal process of quality reviews. Some may spot-check or occasionally critique designs and/or technical decisions. Rarely, designers establish a formal quality management program. These programs are intended to develop checking procedures, checkpoints, lines of communication, meeting processes, clarity of authority and responsibility; assign checking responsibilities to staff; and implement training programs and a wide range of other systems.

Under the pressure of meeting deadlines and maintaining budgets, most designers tend to minimize their quality management processes or to ignore them altogether. Even more distressing is the lack of attention to the development of a program to improve the quality of the product being issued by the firm.

The focus on quality must permeate the entire practice. Quality management is far more than simply reviewing designs or drawings at various stages of completion. It must include not only a review of the product, but also an examination of the method of operation and organization within the firm to develop designs, working drawings, and specifications.

Total Quality Management

A number of years ago, a retired architect explained the difference between quality control and quality management. He said that quality control required three steps:

1. Make a lot of errors
2. Spend a huge amount of time and money finding them
3. Invest even more in fixing them (or ignore them and hope they go away)

He went on to say that quality management was simply a process of avoiding errors in the first place. Many of us today know this as *Total Quality Management (TQM)*.

Unfortunately, TQM has been denigrated by many in the construction industry. Perhaps this was a result of overselling by management consultants, or because of confusion arising from ISO 9000 certification (see the J. Ross Web Added Value material at www.jrosspub.com/wav for more on ISO certification). Whatever the reason, TQM is a sound idea that should be incorporated in all design and construction organizations. While there are numerous philosophical and theoretical works on the subject of TQM, what is important is the practical application of theory. Here are some suggestions on how to apply TQM in your organization:

1. An effective project management/project delivery system is essential to controlling the quality of the product you produce. In nearly every other system, crisis management prevails, a clear point of contact is lacking for clients, no single individual is in charge of a particular project, and a variety of other weaknesses exist.
2. Make the appropriate use of technology. Your manual systems must be well-thought-out and clearly understood before automation occurs. Look toward system integration and continually seek out new applications for technology.

3. Effective communication is the basic tool for improving quality. Accurate and timely information must be conveyed to all who need it. A regular meeting process must be used. Tools such as email, voice mail, and faxes must be part of every PM's daily life.
4. Project management and quality are team concepts. No single individual is capable of doing everything other than on a small project. Every team member must always think about how they can help other members do their job more effectively.
5. Cross training is vital. This allows team members to help each other, builds support during busy times or absences, and improves everyone's performance.
6. Every staff member's authority level must be roughly equal to their level of responsibility. This is particularly true for PMs. Without this equality, staff members will not be able to properly perform their jobs. Unfortunately, many senior managers find it irresistible to meddle in projects, particularly where they are well acquainted with the client or the project. This can destroy the PM's authority and hurt project quality.
7. Every organization must strive to push decision making to the lowest effective level. This is achieved through training, well-documented procedures, good communications, effective hiring, and a variety of other techniques. Remember the three steps to making this work: (1) your system must permit people to make decisions, (2) those people must be willing to make decisions, and (3) they must make correct decisions.
8. *If it isn't broken, don't fix it* is absurd. Continuous improvement is vital. Your marketplace changes daily, the economy changes, legislation and regulations change, the labor market changes, and a thousand other things change. If you do not change—you're finished.

DEVELOPING A QUALITY ASSURANCE PROGRAM*

A QA program must address the quality of all services undertaken by a firm. Why should a firm take the time and make the effort to have a QA program when its staff is perceived to be doing a good job? This is especially true, since the time spent in QA is essentially non-billable. These

---

*The material in this section was prepared by Jeff Orlove, AIA

are valid concerns; however, the benefits are clear and lay in four main areas:

1. **Reduced Liability:** Despite the use of great care, people do make mistakes. Anything that can be done to prevent problems and reduce potential legal claims is definitely worthwhile.
2. **Improved Schedules:** Standardized methods and procedures, and the more organized approaches that a QA program requires, usually make the work proceed faster, resulting in improved schedules.
3. **Fewer Errors and Omissions:** Standardized methods and procedures also result in fewer mistakes. Less experienced staff can use tried and tested information developed by more experienced staff, resulting in fewer errors or omissions.
4. **Higher Profits:** Performing work on a tighter schedule, producing work with fewer errors, preventing lawsuits, and cutting the time required to prepare a defense all result in higher profits.

A QA program can also help design firms in their marketing to existing and potential clients. The program can help assure clients that their projects will be done on time and with fewer errors. A QA program must include more than the construction document phase of a project. It should permeate all areas of a design practice. This includes business development, marketing, accounting, and personnel, as well as all phases of project management. A well-managed office has each of these areas clearly defined, organized, and coordinated. This allows an individual to concentrate on his or her particular area with the knowledge that other activities are being properly managed.

Steps in Developing a Quality Assurance Program

Organization Plan

The first step in developing a QA program is to create a clear and concise organization plan that should delineate the various areas of the design practice. It should also clarify who in the firm has the lead responsibility for each area and how key individuals delegate other activities. In smaller offices, one person may hold more than one position and it is extremely important to clarify roles. The plan must take into consideration the strengths and weaknesses of various individuals and the needs of the firm. Both corporate and project responsibilities must be clearly identified.

Project Manager System

A system for producing projects from inception to completion should be designed. Specific projects should be organized around a PM. The PM is responsible for the schedule, scope of services, fees, and for the organization of the team responsible for producing the design and documentation of the project.

The PM interfaces with design team members, the production and field teams, and with various project consultants. He or she has QA responsibility for that specific project. Whether a project is large and complicated or small and simple, the same basic project management system is used. The only variable lies in the number of team members assigned to a project. The PM may not actually conduct a QA review. He or she must plan for a review in the project budget and schedule and ensure that the appropriate reviews take place.

Quality Assurance Development

One major Chicago-based architectural firm has a principal in charge of QA for both corporate and project activities. This individual works with a project QA director and a corporate QA director. The project QA director develops standardized forms, formats, methods, and procedures to be used by the various project teams. The principal in charge and the QA project director research those methods and procedures they believe would be most beneficial. They also distribute material to the staff and conduct meetings and seminars to disseminate information that will allow projects to be done accurately and on time.

The corporate director of QA deals with similar standardization of forms, formats, methods, and procedures related to general operations. This part of the program includes the following: development of a project control computer system for monitoring fee budgets for specific projects, computerization of the firm's accounting system, organization of management information systems (e.g., project fee, performance, and cost database), creation and maintenance of a technical library for use by the staff, creation of a filing system for drawings, specifications, completed project files, and development of a procedures manual for standardization of frequently performed tasks.

A Lawyer as a Team Member

A very important aspect of a QA program is the concept that prevention is better than correction. Key staff members should be encouraged to consult with attorneys on *what-if* situations so that contract language can be inserted to eliminate problems. This can also help in the drafting of documentation letters to prevent a problem at a future date. Our lawyers have helped develop standardized contract forms. An attorney can also assure designers in certain situations that a perceived problem is not of concern. This allows the designer to give on certain business decisions without increasing liability. This often fosters greater client satisfaction.

Summary

Organization of staff is the key element in any QA program. Clear responsibilities and the use of standardized forms, formats, methods, and procedures will yield the benefits of reduced liability, improved schedules, fewer errors, and higher profits. It is well worth the effort.

PEER REVIEW

Introduction

One important resource that designers can call upon to improve their quality is to make use of peer review programs. These have been offered by most of the major design profession associations including the National Society of Professional Engineers (NSPE), the American Institute of Architects (AIA), the American Council of Engineering Companies (ACEC), and the Association of Engineering Firms Practicing in the Geosciences (commonly known as ASFE).

In May 1990, the *Construction Specifier* magazine reported on the history, progress, goals, and costs of peer review programs. As the magazine noted at the time, "The ACEC first examined the concept of peer review in 1977. ACEC based much of its program on the first peer review program ever created for design professionals, established in the late 1970s by ASFE. Thirteen years later, five major organizations have endorsed the ACEC program, nearly 400 firms have been reviewed and several types of peer review have been created."

In an informal Birnberg & Associates survey of design firms, few undertook peer reviews. Most were not even aware of the existence of peer review programs. Many perceive peer review as a fad of the 1980s. Indeed, peer reviews are infrequent today. This in no way lessens the value of the concept. Why this lack of knowledge and participation? One reason may be the crisis management approach that prevails in many design firms. Often bogged down in day-to-day project problems, many designers fail to examine alternatives to their current methods or to learn about programs that might help improve their quality.

A second reason may be the cost of peer review programs. According to the *Construction Specifier*, "The firm under review is billed for all review costs, including travel and lodging expenses incurred by the review team, administrative costs and daily honoraria. Some smaller firms find these costs prohibitive, but compared to the costs of hiring professional consultants, the price is small." The trade-off for the costs of the review can be great. Clearly, if the firm implements the advice of the peer reviewers, the quality of designs and drawings can be greatly improved and profitability may be enhanced. In addition, several of the leading professional liability insurance carriers offer benefits for firms undertaking peer reviews.

Components of a Peer Review

Firms that wish to undertake a peer review must select a review team from an extensive list of trained reviewers. All reviewers are registered professionals with at least 15 years of experience, including at least 5 years in company management. There is inherently no restriction on the discipline of the reviewers because their technical knowledge is not as important as their management experience.

In a typical review, the reviewers examine documents and interview selected staff from all levels in the firm. Reviews cover six areas, including (1) overall management, (2) professional development, (3) project management, (4) personnel/human resources, (5) finance, and (6) business development (marketing). Firms wishing to take advantage of peer review programs must contact their appropriate professional society. They will be provided a list of qualified reviewers. Obviously, the scheduling of a review can be complicated and can take at least three months to plan for an on-site visit. The review itself can be conducted and completed quickly.

Peer Review: Help or Heartburn?*

The concept of peer review is controversial in some respects. However, as designers become more vulnerable to claims, as projects become larger and more complex, as contracts become more constrictive and threatening, and as clients want more service in less time, designers should take a close look at the way their practices are functioning. Chicago was among the first cities to establish a peer review program for architects and did so

---

*This material was prepared by John Schlossman, FAIA.

within the framework of the AIA chapter back in 1981. A QA task force was appointed as a means of monitoring firms in an effort to curtail rising insurance premiums. Cook County, of which Chicago is the major part, was especially hard hit with professional liability claims at that time. The task force had the following charges:

- To increase understanding of the causes of design and building failure
- To improve communication among architects and other members of the construction process regarding the causes and prevention of building failures
- To elevate the quality of architectural services
- To enhance the image of the architect as a professional able to meet the expectations of the client
- To reduce the number of claims against architects
- To reduce the cost of professional liability insurance
- To reduce the number and severity of building failures and client dissatisfaction
- To attract more architects into AIA membership

Several meetings were held to discuss various methods of attack, to report on issues, and to plan conferences. Peer review was mentioned as a possibility, and numerous questions arose. For example, what constitutes a peer review? Who will do the reviewing? Will the results be known or kept confidential? How long does the review take? What positive or negative effect might this have on the firm's practice, now or in the future? Initial reluctance was anticipated; however, under the tutelage of Chicago attorney Paul Lurie, a leading force in promoting peer review for design professionals, it was decided that a team of professionally related individuals would conduct the initial review as a pilot program. Through interviews and examination of sample documents and resource material, they would gain an overview of the management and operational processes of the firm. From this information, the team could assess whether a firm was operating in a businesslike fashion and performing work in a generally satisfactory manner. The team would then make suggestions on how the firm could improve and what methods should be considered. While general management and operational procedures were reviewed, financial management and technical documents were not.

In Chicago, the peer review team was composed of two or three architects and an attorney. A nominal fee was charged to cover the attorney's expenses and the expense of any clerical functions for preparation of the report. The firm could reject any reviewer it wished, for whatever reason. A letter of agreement for the review was signed by the firm and a declaration of confidentiality was signed by all reviewers. Results were kept strictly confidential and known only to the firm's managing principal. Reviewers received a preliminary copy for comment, which they subsequently destroyed. With small- and medium-sized firms, the review process took the better part of a day. A larger firm needed a second day.

Several important questions are addressed in peer review. Is the firm organized in an orderly manner to deliver work in an efficient way? Is there proper paperwork and documentation to keep projects on track and out of trouble? What is the firm doing to identify potential problems? Does the firm provide enough challenge, satisfaction, and opportunity for staff members? Are they the right personnel to perform the work?

Prior to a formal interview, a questionnaire is sent to the managing principal outlining areas to be covered and requesting that sample materials be available for reviewers (e.g., a set of working drawings, specifications, including front-end documents; contract forms, project logs, personnel manuals); and miscellaneous data such as meeting memoranda, records of telephone calls, shop drawings, and transmittals.

The reviewers interviewed personnel at all levels within the firm, beginning with the principals and including PMs, senior architects or engineers, specification writers, construction administrators, and other staff. Questions included the following: What methods are used to develop and produce documents? How are documents standardized? How is information filed and retrieved? What are the reporting procedures of individuals? How are personnel interviews and reviews conducted? The reviewers tried to determine the spread of responsibility, personnel competence, and office organization.

Once the initial apprehension was overcome, the reviewers found that everyone wanted to be involved. Staff members were genuinely interested in the process and were curious to know the results. At one firm, all employees requested an informal meeting with the reviewers at the end of the day for a general discussion. They were interested in speaking out and learning how to improve procedures. The review process illustrated to them that the firm's principals were interested in improving the firm. After the report is issued, a meeting is held with the firm's principals and the reviewers to discuss the results. Six months later, a follow-up meeting is held with the reviewers to see if suggestions and recommendations have been implemented and to learn their effect on the firm's operations.

11

PROJECT COST CONTROL/ SPECIFICATIONS/VALUE ENGINEERING

PROJECT COST CONTROL

Introduction

Project cost control has two aspects. First, is the need for a design firm to control internal design costs—this requires careful monitoring of expenditures against fee budget. Some of the material discussed below is a recap of earlier information to emphasize the importance of controlling costs. Second, is the need to estimate and monitor the construction budget—failure to adequately do this may result in exceeding an owner's willingness or ability to pay for constructing the facility.

Controlling Internal Project Costs

Perhaps the most difficult task of a design firm project manager (PM) is staying within the project design fee budget. This is a tough challenge for even the most experienced manager. A PM will need help in meeting this challenge. Many factors will affect his or her ability to meet the design firm's goals. As noted elsewhere in this book, the matrix management/ strong project management system has its weaknesses, but it does provide for an individual to manage and monitor the project from beginning to end. To ensure its proper functioning, the matrix system must provide for an equality of responsibility and authority.

No project management system will function well without a capable staff. A high level of experience, accompanied by individuals who are able to make quick and accurate decisions, will go a long way toward keeping you within your fee budget. The goal is to achieve accurate decision making at the lowest effective level in your organization. Many events occur before you sign a design services contract that can have a significant impact on your profit potential. Specialization in one or a few types of projects allows your staff to become knowledgeable in the particular needs and problems of those projects. Research and programming materials, time, and problems can be reduced.

A poorly prepared design scope of services can leave many questions unanswered. This may result in conflict with clients or require excessive, unbilled change orders to meet the program. Poor scope determination can lead to an inaccurately calculated fee budget. The extra work or change orders required to overcome this problem can be very costly. Some firms compound this error by failing to forward price their work. Contracts that last over a long time period (a year or more), or are likely to be delayed must have an inflation clause. Without this clause, overhead increases and staff raises can eat away at the profit margin built into your project multiplier.

Communicating with Clients

Many design firms hurt their chances at controlling project design costs by failing to adequately communicate with clients. This failure covers a multitude of issues. Not adequately defining a scope of services leaves too many issues open for challenge or question, or may result in additional unpaid requests for services. A disciplined process of recording time and expenses related to change orders is essential. It is important to recognize that it is far easier to consolidate information than it is to segregate after the fact.

Scheduling a client's input is essential to controlling costs. Failure to plan for this input can result in delays in decision making. A key to making a profit on a job is to keep it moving smoothly through the design office. Any delays penalize the bottom line. A regular meeting process with your client allows not only better use of your time, but can also provide a decision-making forum. PMs also have an obligation to keep their clients informed. Communication methods such as change order documentation, meeting minutes, and regular telephone calls all help to inform clients.

Information Systems

Perhaps the most important tool needed by a PM is an information reporting system that allows monitoring of costs against the fee budget. This information should be prepared by computer. Many commercially available computer software packages exist. Rarely should a firm seek to design its own software. Any claims that the commercial packages do not meet the particular record-keeping needs or method of doing business of your firm may indicate an incorrect approach on your part.

Most well-run design firms today collect time sheets daily and use electronic timesheets. This improves the accuracy of information and allows more current updating of project status reports. This also allows an interactive process where the PM can use any web enabled device (tablet, phone, laptop, etc.) to check the current status of a project. No information system is of value if the information collected is not accurate.

To control project costs, PMs must understand the information provided by status reports and know how to take action based on the information. If percentages of completions are used, they must be calculated and posted as accurately as possible.

Outside consultants must also be brought into the process of controlling project costs. If they fail to meet deadlines, arrive at incorrect or incomplete solutions, or if they do not segregate change order information, your efforts will be affected or delayed. Wherever possible, communication processes must be established to assist in working with your consultants. If a project seems to be going over budget, prompt action must be taken. It is vital to catch problems as early as possible. This is especially true if your projects are of short duration where any delays in obtaining status reports can prevent effective, corrective action. Staff may need to be changed in order to quickly complete the work or to correct mistakes. Time schedules and budgets may need to be revised to reflect the reality of delays or budget slippage. And, the scope of services must be reexamined to ensure that you are providing what you agreed to do.

Estimating and Controlling Construction Costs

Projects must be managed in a manner that allows for the control of all expenditures. The following examples of estimating construction costs are used with the help of data gathered and rules of thumb. When a quick estimate is required, these methods should serve adequately, but ultimately, more definitive methods must be used.

1. From past projects, cost is divided by the gross building square footage or square meters to determine the cost per square foot or square meter. In order to determine the new building budget, the cost per square foot is multiplied by the gross square footage.
2. Another method similar to No. 1 above, but more specific, is to use past data gathered for individual building types.
3. A third method that is more specific than Nos. 1 and 2 is to use past data pertaining to each trade to determine costs.
4. The method of determining cubic footage or meter costs in lieu of square footage or meter costs has its advantage in projects with large gross volume areas, such as theaters and auditoriums.
5. Other rules of thumb for quick estimation of project costs are cost per unit (material), cost per bed for hospitals, and cost per student for educational facilities.

Various methods that offer more sophisticated results than the rule of thumb methods are available for use by the cost estimator. All of these methods are dependent upon historical data and, obviously, the more current and detailed these data are, the more reliable the estimate will be. Some of the methods used are as follows:

1. Building unit estimating (based on unit costs of material and labor)
2. Statistical and analytical estimating (based on trends, mathematics, and the use of graphs and an overwhelming amount of information input)
3. Quantity survey estimating (based on the determination of the quantity of materials and the amount of time needed to complete specific parts of the construction)

Some methods lend themselves to earlier phases of a project while others are required when a more detailed, concise result is needed. The estimator must have several methods at his or her disposal and must be able to determine which method is most applicable to both the type of project and the particular phase of that project in which the estimate is required. Most of the costs of labor and material information are acquired from suppliers, contractors, and all of the other price determining sources where costs are initiated. These data may be presented directly to the estimator or by way of publications that assemble these data for the estimator who subscribes.

Many publishers of periodicals and magazines offer various types of construction cost information. The estimator's good judgment is ultimately the determining factor as to whether or not the ongoing generation of cost analysis is maintained as accurately as possible. The human factor is not replaceable. Human error, on the other hand, can be somewhat eliminated by the use of computers, which not only calculate costs and analyze results, but also store cost data for use in determining construction costs. There are additional factors that cause cost differentials in building projects and these factors must be considered. They are the elements of solution for one portion of the building against those decisions made for other parts (or systems) of the building. It involves an overview of all parts of the project and the evaluation of all implications of a design solution.

SPECIFICATIONS

Introduction

Specifications and construction drawings serve complementary functions. The drawings show the where, what, and quantity information about the materials and equipment; while the specifications show the quality of materials and equipment, the installation methods and techniques, and the results to be achieved. Construction drawings and specifications are generally developed concurrently and all changes must be accounted for in both documents.

Variations will be found between design firms as to the completeness and quality of the construction drawings. Variations occur in document preparation even within the same firm due to limitations of the architect's or engineer's design (not construction) budget, variation in control between PMs and job captains, and so on. Some firms have high quality management standards that require detailed drawings despite budgets. Others place a greater reliance upon either more detailed specifications and/or upon the general contractor.

Generally, the specifications set the quality of materials and equipment, the installation methods and techniques, and the results to be achieved. Specifications are included in a document called the *project manual*—the terms are often used synonymously. Specifications may be prepared on a full-time basis by one or more individuals in large firms, and by someone who has a variety of other responsibilities in smaller firms. With the enormous increase in the number of available products and the increased complexity of building construction techniques in recent years, it has become virtually impossible for any one individual to keep abreast of all developments in the industry.

On most projects, large parts of the specification may be taken from prior jobs as a time-saver—this is usually adequate. Most firms of any size have a master specification file maintained and filed by product subject, and nearly all design firms have this file computerized. Clearly, computerization increases speed and accuracy. Standard specifications are also available from associations and industry, such as those prepared by the American Society for Testing and Materials (ASTM).

It is the responsibility of the design professional to select materials and equipment. Six basic approaches are used in the specifying of products (see Figure 11.1). With the use of trade and brand names and the ready availability of substitutes, in many cases this is not an easy process. The six approaches are as follows:

1. **Non-competitive or closed specification:** This type of specification lists only one trade name for each product required. The contractor is required to furnish this product without substitution. This approach is rarely permitted in public work.
2. **Contractor's option specification:** This specification lists every trade name that will be accepted for each product. The contractor may use any product listed at his or her option without consulting the designer, but unlisted products will not be allowed.
3. **Product approval specification:** This type lists one or more trade names for each product with a statement that any requests for substitutions by a bidder (before the award of the contract) be sub-

---

1. Non-competitive/Closed Specification
2. Contractor's Option Specification
3. Product Approval Specification
4. Substitute Bid Specification
5. "Or Approved Equal" Specification (Open Specification)
6. Product Description or Performance Specification (Open Specification)

Figure 11.1 Selection of products

mitted in advance. The advantage of this system is that products of which the design professional was unaware can be brought to his or her attention and can be included in the list of products allowed.
4. **Substitute bid specification:** This specification lists one or more trade names for each product just as the product approval specification does, but the bidder's base bid must be compiled using one of the listed products. The bidder may, however, attach to his or her bid a proposal for substitute materials and show the amount to be added or deducted from the base bid if the substitution is accepted. The advantage of this approach is that the contractor is afforded the opportunity to bid on materials he or she prefers to use and then a choice can be made weighing the value against the extra cost or savings.
5. **Or approved equal specification:** This type of specification lists one or more trade names for each required product and follows with the phrase *or approved equal* or there may be a statement that these words are to be implied whenever a trade name is used. This is considered an *open specification*.
6. **Product description or performance specification:** This specification describes completely all details, qualities, functions, and sizes of product without mentioning a trade or brand name. Any product that meets all the detailed requirements will be approved. This type is often used in public work and is also considered an *open specification*.

Contracts often require the contractors to submit samples to the designer for approval where indicated in the specifications. This is to ensure that the material is meeting the standard specified. Products do not always reach the project site as a result of being initially specified. Often, addenda are issued by the architect who has several review and approval options available. As noted earlier, there is sometimes a pre-bid substitution of products by the general contractor (product approval). Post-bid substitutions may also be permitted by the supplementary conditions to the construction contract. Figure 11.2 summarizes and compares various types of specifications.

Other Specification Issues

Allowances

Allowances are used in lieu of a full specification of a product. They are used for deferred product selection by the architect. Examples would

	Base Bid	Modified Base Bid	Base Bid with Requested Alternatives	Base Bid with Alternatives and Substitutions	Or Equal	Modified or Equal
	(Proprietary 1 Product)	(Proprietary 2 Product)	(Bids based on base bid but alternatives may be listed.)	(Bids based on base bid but quotes on alternatives will be accepted or reduced costs credits.)	(Specified products min. qualities must be met.)	(Spec. products min. quality must be met but submission of unspecified products will be considered.)
Contract influences product selection	No	No	No	Yes	Yes	Yes
Permits early detailing of special features and quick completion of working drawings	Yes	No	Yes	Yes, but changes likely	No	No
Allows exact calculations of dimensions and simplification of estimating	Yes	No	Yes, but less than Base Bid	No	No	No
Owner has early indication of complete project quality	Yes	No	Yes	No	No	No

Source: *Building Products Marketing Manual*, by Howard Birnberg and Leonard Robin

Figure 11.2 Specification types table of comparisons

be in the selection of hardware, artwork, or other subcontract work. The allowance usually covers the net cost of purchase of the product, with taxes paid and delivery to the site. Installation and on-site handling are generally not included. Unit prices are used when there is doubt as to the quantities required. A request for unit prices in the original bid removes the need to negotiate these items for price when the eventual quantity is established. Unit prices should include material, installation costs, overhead, other expenses, and the contractor's profit. Unit prices are frequently offered for both increases and decreases in quantity.

Performance-Based Codes and Standards

Traditional building codes and standards have been prescriptive in nature, describing an acceptable solution. These codes prescribe exactly the materials to be used in an application. They are normally easy for a builder to understand, relatively simple for a third party (such as a building department official) to check, and reasonably simple for an inspector to enforce. However, two significant problems occur in the use of prescriptive codes: (1) they often serve as a barrier to innovation and (2) they make it difficult to use more cost-effective approaches to construction. An alternative is the use of performance-based codes and standards. This alternative specifies desired results such as the ability to withstand fire for a stated period of time without prescribing the exact materials to be used. The International Council for Research and Innovation in Building and Construction (CIB) defines this as, "The performance approach is, in essence, the practice of thinking and working in terms of ends rather than means." Rather than prescribing how a building or product is to be constructed, performance codes and standards specify what it is to do.

VALUE ENGINEERING*

Introduction

The initial impetus for value engineering (VE) in the design and construction industry came from U.S. federal government agencies in the 1960s. Fifty years ago, VE was viewed as a way of reducing over-budget project costs. The old VE focused on cost reduction, frequently at the expense of client desired project features. Today, a VE study seeks to balance client required project performance with the costs to achieve it.

---

*This section prepared by Howard Ellegant, AIA.

This changed focus allows PMs to offer a separately identifiable service and involve client and building team representatives in an intimate, exciting, and time-limited process. This builds team camaraderie, deepens project understanding, and opens communications between project team members.

Modern VE fits right into the proactive practices of Partnering and Total Quality Management (TQM) by strengthening project team relationships, empowering decision making at the project team level, and identifying criteria by which customers (project owners and users) will judge the success of the finished project. A successful VE study has four principal characteristics:

1. Implemented results
2. Customer involvement
3. Thorough function analysis
4. Strict adherence to the VE job plan

Implementation Results

Implemented VE team recommendations are a primary measure of VE success. To ensure full benefit of team recommendations, the owner and designer must buy into them. The surest way of achieving this is to actively involve them in the VE process. To *ROAR* down the implementation path requires *R*esponsibility for implementation, *O*wnership of the recommendations, *A*uthority and *R*esources (ROAR) to implement the VE study team's recommendations. An empowered project team can have all of these ingredients, and it is the PM who coordinates them. The biggest stumbling block to roaring ahead with implementation is lack of ownership of the recommendations.

Still practiced today, the old VE uses a peer review team to perform an objective VE study of the design team's work. An immediate consequence is to establish an adversarial relationship between the design team and the VE study team. The owner is caught in the middle and must decide between the VE team's recommendations arrived at over a short period of time and the design team's often defensive reaction to them. The very people who have to approve and implement the recommendations have no ownership of them and no stake in a positive outcome!

People do not like change, but if they participate in creating it they understand and own it—and can ROAR their approval and acceptance! Using the project team as the VE study team dramatically increases implementation by adding the vital factor of ownership to responsibility, authority, and resources. The VE process, specifically through function analysis facilitated by a knowledgeable VE specialist, provides the necessary objectivity for project review.

Involve the Customers

The old VE too often focuses on the *engineering* and does not pay enough attention to the *value*. I define *value* as the relationship between customer project acceptance and project cost (value = acceptance/cost). Acceptance, in turn, is composed of customer expectations and how well the project meets them. Cost includes initial capital plus follow-on costs—total life cycle costs. While designers influence costs through their responses to customers' requirements, the customer is the final judge of good or poor value of the result.

The old VE concentrates on driving project costs down through cost-saving recommendations to change scope, materials, and systems—often with little or no regard for owner acceptance. Modern VE also suggests cost saving changes, but does so within a context of customer project acceptance criteria and their relative importance to each other. Seeking a balance of project performance and costs results in cost savings as well as reinvesting some of the savings to increase project acceptance.

At the start of a VE study, we conduct a focus group to solicit from the customers (owner, users, and operators) their criteria for a successful project. The VE team uses this information to determine where to concentrate its effort and generate alternative solutions to improve project performance and reduce costs by correcting problems the customers highlight.

Modern VE uses a multidisciplinary team of project stakeholders to break down the design into functional performance elements. Costs and benefits are assigned to each element. The value of each functional performance element is measured by comparing its costs and benefits. Total project value is improved by making appropriate changes to balance performance and cost. Owners and designers play an integral role in defining project performance, and generating and evaluating alternative ideas based upon the technical feasibility, cost impact, and political acceptability of each idea. Modern VE gives the PM the opportunity and the tools to manage both project cost and owner satisfaction.

Function Analysis

The process of breaking the design down into performance elements is called *function analysis*. It is a systematic approach to identifying and analyzing what customers need and want from facilities to support their operations. Function analysis is the foundation of VE, and it is crucial to a successful study that proposes changes to improve project value, not just reduce cost—changes that do not detract from performance or aesthetics but enhance both.

As a purchasing agent during World War II, Lawrence D. Miles, VE's originator, specified functions to suppliers and let them suggest what materials, parts, or assemblies could be used to do a job instead of the ones specified by the engineers, many of which could not be procured at any price because of wartime scarcity. Miles was using performance specifications, but these were performance specifications with a difference. They focused on the reason that a part existed rather than its longevity, strength, elasticity, or other engineering characteristics. Miles's performance characteristics (functions) were identified by understanding how the part would be used—the job it had to do for the customer. Miles also reasoned that the cost of accomplishing functions could be derived by careful analysis of the contribution particular materials and manufacturing operations made. He also believed in the assignment of incremental costs of materials and manufacturing operations for a given part to the function that it provided. A portion of the cost of galvanizing a part might be assigned to the function *resist corrosion*, another portion to *improve appearance*, and the cost of drilling holes to the function *allow attachment*. (If a slotted hole, then the cost to elongate the hole would be attributed to a function such as *adjust alignment*.) The VE study team identified alternative ways to perform the various functions, usually concentrating on those with the highest cost.

In the early 1960s, Charles W. Bytheway created the function analysis system technique (FAST), a structure within which (and rules by which) to logically diagram the relationship of Miles's functions. In the 1970s, Professor Thomas J. Snodgrass of the University of Wisconsin contributed a system to collect user and customer attitudes about VE study object and rate their importance, assign the importance ratings to functions, and then compare the importance of a function to its cost to identify a value mismatch—a function for which the cost, whether high or low, was mismatched with the customer's perception of its importance. The team concentrated on improving the mismatched functions.

Recently, I led a VE study for a new county jail in California. A significant amount of floor area was consumed by a property storage room. Construction costs allocated to the functional performance requirements of receive inmates, classify inmates, and release inmates totaled over $3 million—10.5 percent of estimated construction costs. Armed with this information, the jail commander decided to eliminate inmate property storage in this new detention-only facility since the jail's primary function was to manage sentenced inmates. Inmate release and new inmate intake will be done at the existing jail and inmate property will be stored there. The first cost savings were $500,000. Life cycle cost savings from eliminating the property storage center staff position and avoiding property storage equipment maintenance were another $1.3 million. Function analysis had provided the rationale for a decision that up until the time of the study was only a distant thought.

A thorough function analysis reviews closely the relationship of three things: (1) function, (2) function cost, and (3) function importance or worth to the customer. The result reveals targets of opportunity for overall project value improvement, not just cost reduction. This core VE technique can be used at any project stage from site selection through construction, to help the project team identify and clarify essential project performance requirements. Studies very early in project conception may not require a rigorous assignment of project costs to functions. The process of defining functions and building a function diagram builds understanding of the project.

During a one-day study, we helped the CEO and top executives of a health care plan management company use function analysis to review, evaluate, and decide between three existing buildings being considered as new locations for the company. While building the function diagram, a breakthrough occurred when the team agreed that their single reason for moving was not to improve operations, but to increase sales by providing the firm with a better image to prospective clients who visit their offices.

After reviewing the three potential locations in this new light, the team concluded that none were appropriate to their main requirement for moving. They sought other buildings which could present an image to prospective clients that better reflected their professionalism and capabilities. The design architect was part of the study. The completed function diagram clarified and documented for the architect the client's project performance criteria for the design of the new offices. Assigning functions to activities (e.g., claims processing, reception, customer service) was the initial step in developing design criteria for the spaces in which these activities would occur.

Function analysis and function diagramming as a stand-alone technique afford the PM a powerful consensus building and analytical tool during project planning. During design, assigning costs to functions puts a price tag on program decisions allowing the customer to answer the question, "Am I getting the result I want for the money I am spending?"

Adhere to the Value Engineering Job Plan

Ready, fire, aim! is not the most effective sequence if the mission is to hit the target with maximum effect using the minimum amount of ammunition. The VE job plan (see Figure 11.3) establishes a sequential order of activities, in measured blocks of time, for the VE study team to follow. Each block contains specific tasks to be completed leading toward a desired result. The time allotted to each job plan step depends upon the scope of the project studied.

Following the VE job plan gives the PM the control necessary to limit the duration of the study while ensuring successful results. The job plan is a flexible tool whose order must be preserved, but its duration can be extended or compressed to fit the project scope. The activities within each step also vary depending upon when in project development the study is carried out. I usually recommend a three-day team study at the pre-design phase. A longer study, up to six or seven days of team meetings, with

Pre-Study
Review project
Review costs
Select team
and site

↓

Information
Customer attitudes
Perform function analysis

→

Create
Generate and champion ideas

→

Evaluation
Select ideas, outside vendors

→

Decision
Choose ideas to develop and present in report

→

Presentation
Submit report

↓

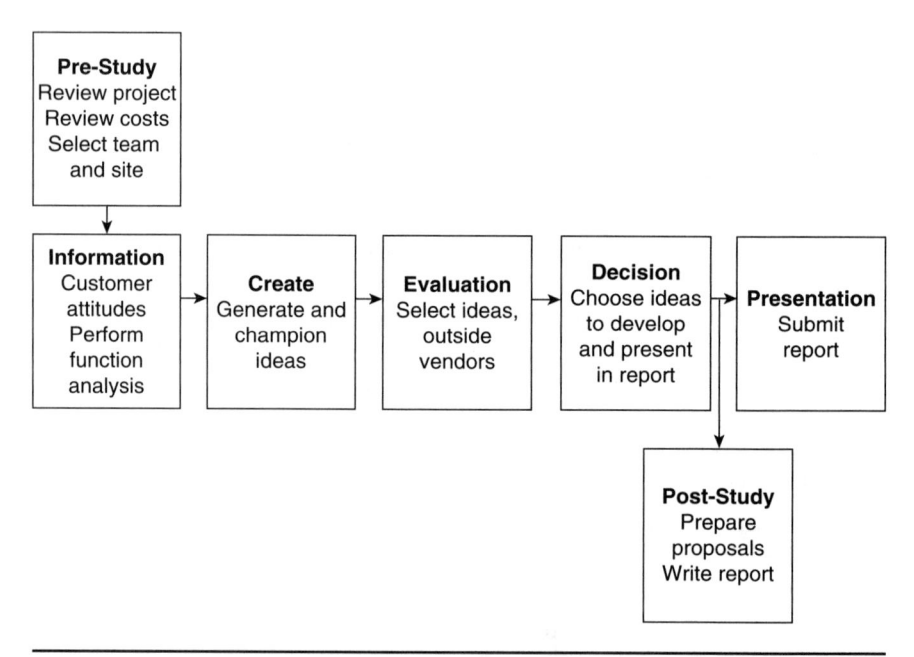

Post-Study
Prepare proposals
Write report

Figure 11.3 Value engineering job plan. (*Source*: Howard Ellegant, AIA)

individual team member's assignments to be completed between some of the meetings, is effective during concept design or later.

It is normal to think that the VE team spends most of its time in the creative phase, generating alternative ideas. In reality, the creative phase is the shortest phase of the VE job plan. The most time-consuming and most important VE study activity is a thorough function analysis performed in the information phase. I compare the time the VE team spends performing function analysis with the time my son spent in high school doing his homework. He learned there was a direct correlation between the time spent on his homework and increasing his grade point average.

The information phase is the VE team's homework, and function analysis is the primary lesson. In the two to three days the team may take to work through it, team members become primed, collectively and individually, to generate by function hundreds of ideas in a short period of time. As a facilitator, my major problem can be halting the creative phase. For example, in a training workshop last year, the lowest number of ideas generated by a team was 198, while the highest was nearly 300—in only four hours! This profusion of ideas are the seeds from which project value enhancing changes are grown through thoughtful evaluation, careful development, and owner buy-in.

Conclusion

Modern VE employs all of the characteristics of proactive, customer-driven processes increasingly being used in project design. A multidisciplinary team of stakeholders working together to understand and accomplish project objectives is the foundation of Partnering. Thorough function analysis identifies customer quality objectives and the cost to achieve them. It helps the customer articulate required project performance and an acceptable cost to achieve it, and then concentrates the entire project team's energy on making it happen. The VE job plan is a framework within which to conduct meetings for the project team to focus on identifying the project mission, objectives, and performance and then working together to achieve them. Modern VE is a powerful tool to help the PM work with the client and the project team to achieve a quality project at affordable costs.

12

SCHEDULING

PROJECT SCHEDULING

Introduction

Project schedules are important and highly useful tools for designers and contractors. The preparation of schedules allows for detailed planning of work activities and provides a device for communicating critical dates and activities to clients, consultants, and the internal project team. In addition, schedules provide a tool to help program the project, test alternative approaches, and evaluate job performance.

The development of a project schedule is an aid—not a substitute for project management. It is a useful method for monitoring percent complete, offers guideposts for the project team, and can show the likelihood of meeting deadlines. The first step in developing a project schedule begins with defining the planning units. This may be based on geography (locations), function (e.g., architect, engineer), or phase (e.g., design, construction documents). Secondly, decisions must be made as to the kinds of information a schedule should show. This may simply be a list of activities to be accomplished or may include time frames, sequences, and individuals responsible.

Scheduling Methods

An appropriate scheduling method must be selected and prepared in written form. While many computer programs are available to accomplish this step, care must be exercised not to make the schedule overly complicated. As a communication tool, it must be presented in a clear manner to achieve its primary purpose. Many scheduling methods exist and each has its own strengths and weaknesses. Several are included here.

Full Wall Scheduling

This is an antiquated method by which project tasks are listed—the individual responsible for each task is noted and a preliminary schedule prepared. The tasks are listed on 3 × 5 index cards and divided into piles for each responsible party. The list of individuals is posted on one side of an office wall and the time frame/periods listed along the top of the wall. All participants are assembled and the tasks are tacked to the wall based on when the activity will be started and finished. In this method, a schedule is developed and all participants understand their own responsibilities.

The disadvantages of this method are many. To be most effective, all parties (including the client, consultants, and contractor) must be available to participate in the planning session. On large projects, this can be an extremely time-consuming process. An advantage is that it allows a high degree of interaction between all project participants at an early stage of the job.

Bar Charts

This is a time-tested and familiar device for scheduling. A bar chart shows a list of activities, the responsible party, and the duration of an activity. Bar charts, also called *Gantt charts*, are easy to prepare, are familiar to most people, and clearly communicate information (see Figure 12.1). In addition, they are excellent and simple tools for monitoring. Other methods, however, are superior for planning purposes. Bar charts do have several disadvantages. They do not show sequencing of events/activities and the interrelationship among tasks. In addition, on a bar chart every task appears equal in importance.

*Critical Path Method**

The critical path method is perhaps the most widely used scheduling method. The network schedule has its ancestry in the bar chart. The inadequacies of this method resulted, in 1956, in motivating the DuPont

---

*This material on the critical path method of scheduling was prepared by Thomas Eyerman, FAIA, and is taken from his "Financial Management Concepts and Techniques for the Architect" (1973).

	Months							
	1	2	3	4	5	6	7	8
Design A & B								
Construction A & B								
Installation A & B								

Figure 12.1 Gantt chart

Company to adopt the rapidly growing power of the computer to construction scheduling. The system they developed is called the *critical path method* (CPM). At about the same time, the U.S. Navy developed a system called *program evaluation and review technique* (PERT). The primary difference between the two is that CPM uses one time estimate and PERT uses three (most likely, optimistic, and pessimistic).

Many CPM and PERT applications in the construction process occur during the actual construction phase. The networks that are developed can range from less than 100 to many thousands of activities. It is the development of a disciplined method of planning sequences that is important in CPM and PERT, not simply the use of the computer. Not only is the planning process significant with this method, but it also encourages the effective management of schedules.

In developing a CPM schedule, the project manager (PM) must identify the interrelationships between tasks and establish their duration. He or she must prepare a schedule of activities and determine the critical paths or steps. The first and most important step in preparing a project for networking is the correct division of the job into units of work that are relevant to the people for whom it is being prepared. For example, a network indicating just three subjects—design, working drawings, and construction—may be applicable for the design professional to show a client. For internal use, however, each of these subjects might be subdivided—such as working drawings—into architectural, mechanical, and structural drawings.

Furthermore, to be of any aid to technical staff, the drawings might be subdivided into plans, sections, and elevations. Networks for internal control must be developed at the level where the work is to be performed. Once the network process is removed from the operating level, it becomes merely a theoretical tool. No matter how detailed and accurate a network may be, if it is not meaningful to operating staff, it is useless.

The personal involvement of these individuals is what makes a network a viable tool. A network has three phases:

1. **Planning:** This is the study or recognition of which events and their interrelationships are necessary to complete a project. A network is a systematic attempt to plan the work. Just like a wall section shows how a wall is to be built, the network is a graphic representation of how the project will be produced in the office. The people performing this task must have a sound knowledge of how a job is put together. Moreover, they must have a thorough understanding of the particular project and of the scope involved.
2. **Scheduling:** This is the second phase of networking. A reasonable estimate is made of the time required to perform each event shown on the network developed in the planning stage. With the network and the estimated time for each activity, you can proceed in an entirely mechanical manner to determine the overall time required for a project. By establishing the time required, the design professional is establishing the budget for the project. Just like dimensions on a wall section tell how high the structure will be, the schedule tells how long it will take to produce the project.
3. **Monitoring:** This is the third phase of networking. Monitoring means nothing more than comparing what is actually being done with what was planned. In other words, *we have planned the work, now we work the plan.* The network enables you to spot difficulties at an early stage of the project. This is the point where the accounting reports must tie in with the network.

The level at which the designer or contractor requires a network—commonly called the *level of indenture*—is determined by *who is it for?* and *what is its purpose?* Answers to these two questions are the first management decisions in developing a network. Once the level of indenture has been established, each division of work is split into a number of component parts called *activities*. An activity is any definable time-consuming task necessary to execute a project. An example of an activity would be *drawing wall sections*.

The start and completion of an activity is called an *event*. An event is a specific point in time indicating the beginning or ending of one or more activities. An example of an event would be *wall sections completed*. An example of an activity would be *preparing wall section of exterior wall*. In preparing a project for networking, it is most useful to define the last event and then work backward from that point. As the activities and events are developed, they are plotted on paper to create a pictorial description of the network. An arrow represents an activity and a circle represents an event. PERT is generally event-oriented and CPM is activity-oriented.

Events and activities must be sequenced on the network under a logical set of ground rules that allow the determination of critical paths. These ground rules are found in Figure 12.2. In Example 5 of the figure, X is called a *dummy variable*. It represents a restraint (halting start of activity C until activity B is completed) that cannot be recognized by the conventional symbols for events and activities.

How can the networking technique represent activities that are shown in a Gantt (bar) chart as illustrated in Figure 12.1? The answer is that there must be something that causes the decision to start an activity. The task of the person developing the network is to isolate this decision-causing point and to include it as an event on the network. In the preceding example, the decision to start construction of A is determined by approval of the design for A. Figure 12.3 shows a sample network for a solution of the problem. When the network is completed, an estimate is made as to how long each particular activity will take.

A further development in networking can be to calculate the earliest and latest time an event can take place without affecting the completion

No.	Example	Rule
1.	A ──○── B	Activity A must be completed before activity B can start.
2.	A, B ──○── C	Activity C cannot start until both activities A and B are completed.
3.	A ──○── B, C	Activities B and C cannot start until activity A is completed.
4.	A, B ──○── C, D	Activities C and D cannot start until both activities A and B are completed.
5.	A ──○── C; B ──○── D with X dummy	Activities A and B must be completed before activity C can start. However, only activity B must be completed before activity D can start.

Figure 12.2 Activity ground rules. (*Source:* Tom Eyerman, FAIA)

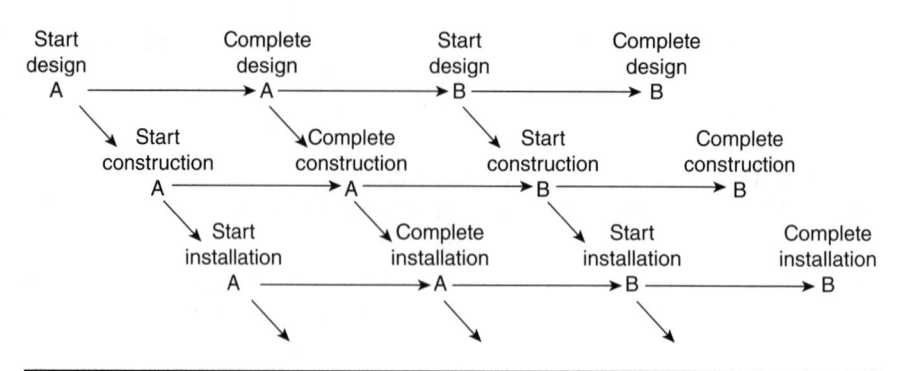

Figure 12.3 Sample network. (*Source*: Tom Eyerman, FAIA)

of the project's final event. The difference between the earliest and latest time an event could occur is called *slack time*. The critical path is then simply the series of activities and events that have no slack time. The critical path, therefore, is the bottleneck route. Only by finding ways to shorten jobs along the critical path can the overall project time be reduced; the time required to perform noncritical jobs is irrelevant from the viewpoint of total project time. The frequent (and costly) practice of *crashing* all jobs in a project to reduce total project time is thus unnecessary. Of course, if some way is found to shorten one or more of the critical jobs, then not only will the whole project time be shortened, but the critical path itself may shift and some previously noncritical jobs may become critical.

Once the network is developed and the critical path established, cost estimates are made by first determining the necessary personnel to perform each activity. The personnel allocation is then converted to dollars to determine the direct cost of the activity. These cost estimates are used in three ways:

1. To aid the professional in determining his fee
2. To evaluate the project in terms of expenses before any action is taken
3. To provide benchmarks against which actual costs can be compared

This then describes the networking technique. A single example should help to clarify the process of constructing a network. The example project of building a house is taken from the article, *ABC's of the Critical Path*

Method (Levy, Thompson, and West, 1963). Shown in Figure 12.4 is a list of major tasks together with estimated time and the immediate predecessors for each task. Following the ground rules for networking, a network can be developed as shown in Figure 12.5. The path A-B-C-D-J-K-L-N-T-U-X is the critical path, with a maximum of 34 days.

Suppose now that October 1 is the target time for completing the project. This October 1 date is subtracted from the time for Event X and the remaining time is forwarded to Event S. Assuming a six-day work

Task No.	Immediate Predecessors and Description		Normal Time (Days)
A		Start	0
B	a	Excavate and Pour Footers	4
C	b	Pour Concrete Foundation	2
D	c	Erect Wooden Frame Including Rough Roof	4
E	d	Lay Brickwork	6
F	c	Install Basement Drains and Plumbing	1
G	f	Pour Basement Floor	2
H	f	Install Rough Plumbing	3
I	d	Install Rough Wiring	2
J	d, g	Install Heating and Ventilating	4
K	i, j, h	Fasten Plaster Board and Plaster (including drying)	10
L	k	Lay Finish Flooring	3
M	l	Install Kitchen Fixtures	1
N	l	Install Finish Plumbing	2
O	l	Finish Carpentry	3
P	e	Finish Roofing and Flashing	2
Q	p	Fasten Gutters and Downspouts	1
R	c	Lay Storm Drains for Rain Water	1
S	o, t	Sand and Varnish Flooring	2
T	m, n	Paint	3
U	t	Finish Electrical Work	1
V	q, r	Finish Grading	2
W	v	Pour Walks and Complete Landscaping	5
X	s, u, w	Finish	0

Figure 12.4 Example network #1. (*Source*: Tom Eyerman, FAIA)

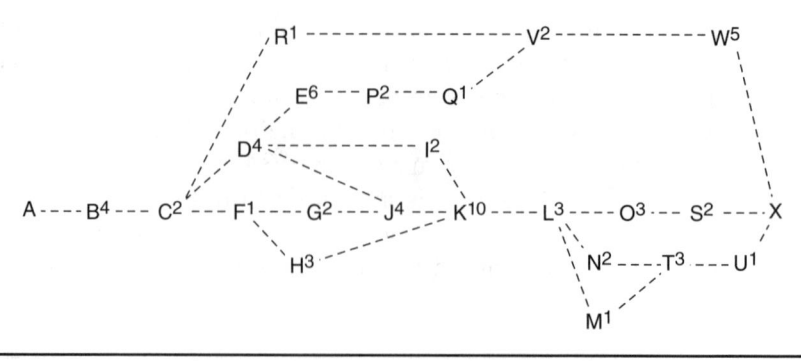

```
        ,R1 ------------------ V2 ---------- W5
       /                    ,'              \
      /  E6 --- P2 ---- Q1'                  \
     /      ,'      ,'   ,'                   \
    /   D4 ≲-----------------I2                \
   /  ,'  ,' ,'         ,' ,.'\                 \
  A----B4---C2---F1----G2----J4---K10----L3----O3---S2----X
                 `,    ,'          `,          ,'
                  H3 -'              N2----T3---U1
                                      \      ,'
                                       M1
```

Figure 12.5 Example network #2. (*Source:* Tom Eyerman, FAIA)

period, you can see that Event A must be started no later than August 23. Using the same analysis, you can see that Event V may start as early as September 13 or as late as September 24.

Benefits and Limitations

The benefits of using a network are as follows:

1. A network gives the designer and contractor a method of programming the project and then a method of evaluating performance during the project, not after the project is completed.
2. A network provides a plan that can be distributed to the project team so they know where they are headed and how they are going to get there. Furthermore, a network, by indicating a percent of completion for various calendar days, provides the designer with a good check on the actual percent of completion. Everyone on the project team knows what must be completed by a certain date. Thus, rather than look at a completion date several months in the future, their attention is drawn to what they must accomplish in the next two-week period.
3. A network is an excellent tool to show current and potential clients what must happen in the office to produce a set of drawings. It may help the client reach a decision if he or she knows the completion date may be extended a month due to his or her indecision. Also a network can show very early in the project whether the scheduled client dates can be made. As a result, the design professional is able to give the client realistic dates of completion, while avoiding the embarrassing task of explaining to an owner why the contract documents will not be finished on time.

4. The network may be used to determine what fee is necessary to perform the work a client demands.
5. A network gives the designer a basis to determine the future personnel requirements for the office. The networks will never give a completely accurate staff projection since there are always jobs requiring staff, but are too small to network. With experience, however, the professional should be able to calculate accurate staffing requirements.
6. A framework is developed that can be used to test alternative approaches to a project.

The benefits described do not just happen. The designer and contractor must become involved with, support, and use the network. But, in order to do this you must have the following:

1. The ability to fully define the project to a point where events and activities are clearly expressed
2. The ability to keep the size and complexity of the network reasonable
3. Cooperation in developing a meaningful network

In summary, networking cannot be delegated; rather it requires active support and judgment of the PM.

There are certain limitations of networking:

1. Networking is not a panacea for ineffective management. It is an improvement over other project planning devices, such as Gantt charts. It also allows for the generation of analytical information that was not previously available.
2. The introduction of networking may increase costs. Consequently, the cost of introducing networking into a specific situation must be evaluated against the increased benefits received from the additional information that may become available. This evaluation will not always favor networking.
3. Networks have caused some people to look backward, seeking to place or shift the blame for a lack of progress rather than look forward to bring the project to a successful conclusion.

For many networking applications, simple personal computer-based software is available. This makes the use and benefits of network scheduling available to all design firms. Numerous other scheduling methods are in common use. The milestone chart shows activities, duration, and start and finish dates. Others include the cumulative graph, histogram, and project chart. All have their uses and application.

ISSUES IN PLANNING AND SCHEDULING FOR PROJECT MANAGERS*

Introduction

Outside of the construction industry, the term project management often broadly means scheduling. Outside, project management software is scheduling software. Inside the industry, this is only one fragment of the project management package. Despite this limited role, project schedules are important and highly useful tools for designers, facilities managers, and contractors. The preparation of schedules allows for detailed planning of work activities and provides a device for communicating critical dates and activities to clients, consultants, designers and all other team members. In addition, schedules provide a tool to help program projects, test alternative approaches, and evaluate job performance.

The development of a project schedule is an aid to, not a substitute for, project management. It is a useful method for monitoring percent complete, offers guideposts for the project team, and can show the likelihood of meeting deadlines. A schedule sets the *when* (such as dates and durations) and provides a time-based *plan* for performing the work. In general, PMs must plan, manage, and control five important aspects of the project. These include: materials, equipment, people, money, and time. This article discusses the importance of project schedules in controlling these resources.

Defining Successful Project Completion

A necessary first step in the preparation of a project schedule is to define what it means to successfully complete a project. While this definition will vary from project to project, owner to owner, and by team members, there are several widely accepted parameters. For PMs whose primary areas of responsibility and authority are scope, schedule, and budget, these parameters should be very familiar.

---

*Part of this material was prepared by Bradford Sims, Ph.D., in addition to the author.

1. **On-time:** There can be a significant cost, particularly to the owner, of a delayed project. This can be in the opportunity cost of not having use of (or access to) a facility, increased cost of construction materials and labor, loss of productivity, the loss of markets and market share, and a wide variety of other penalties resulting from delay in the completion of a project. Designers and owners pay a similar penalty as they must continue to make staff and equipment available on a project generating little or no revenue, they may suffer the potential loss of other projects due to the inability to assign needed staff and equipment, they may experience damage to reputations, etc.
2. **Within budget:** For designers, contractors, and subcontractors, any delay in completing a project can result in significant cost implications. Penalties may be incurred and/or assessed when deadlines are missed; risk may increase as work is rushed to meet deadlines; and direct costs for materials, labor, and equipment continues even as the revenue generated is reduced or ended. Owners are sometimes responsible for slippage in deadlines as scope changes are made late in design or during construction and must be managed. Unexpected issues often arise as materials are late in delivery, products are damaged in shipping, fabrication does not meet requirements, and a myriad of other issues cause problems in meeting budgets. Despite these potential pitfalls, schedules can be of great help in keeping budgets on track.
3. **Meets the owner's needs/program:** Whether it is a building, highway, study, or research report, a project exists to meet the owner's needs/program. Successful projects achieve this goal. The entire design and construction team must constantly keep this in mind and, as one measure, define success as meeting the client's program.

Purpose of Planning

A project plan has several primary purposes. First, it provides a *road map* to show the route(s) to completing a project on-time and within the construction budget. Second, a project plan lays out the sequence and logic of activities on a job. This sets the *who, what, when, where and how* on the project. Third, the plan forces the careful consideration of goals and objectives often in the three categories of optimistic, pessimistic, and realistic. Lastly, project planning has the vital goal of reducing uncertainty by minimizing conflict and confusion among team members.

Planning Activities

When preparing a project plan, there are a number of key planning activities to consider. A careful analysis of the timing of all phases of the work must be examined. The types and the availability of materials to be installed need to be listed and quantities considered. A similar analysis must be prepared for tools and equipment and their availability analyzed. Most important, the personnel needed, the skills required, and the timing of employing these people evaluated. Finally, a careful examination of all aspects of the project should be undertaken to consider areas of potential conflict. Solutions must be evaluated.

Why Schedule?

Scheduling serves many purposes. The communication of the construction plan to all team members is a crucial one. This includes the coordination of activities, deliveries, installation, etc., and outlines the methods, sequences, and timing of the construction. Scheduling plans provide day-to-day goals for measuring performance and accomplishments. In addition, the schedule offers a device to monitor project progress. As scope change is inevitable on a project, the schedule allows a useful method to manage that change. While there are clearly many other benefits to scheduling, these are some of the most significant.

What Constitutes a Good Schedule?

There are three primary components to an effective project schedule. First, it must be realistic, second, it must be carefully planned, and third, the activities planned must be obtainable and achievable based upon the resources available. While a schedule does not *need* to be prepared on a computer, today nearly all schedules are generated with software such as Microsoft Project. The CPM is the most widely used scheduling approach. Whatever method is used, a schedule must be prepared with care in order to be realistic and achievable.

Products of a Schedule

While there are many products resulting from the preparation of a project schedule, the following are some of the most significant.

1. Project schedules allow you to monitor the productivity of your own work force and that of other team members including subcontractors to the general contractor.
2. Time charges and financial expenditures on project activities can be regularly monitored and evaluated.
3. Schedules produce significant information that can be retained in a database for use in preparing future time and money estimates.
4. In the event of disputes, a project schedule is useful in the resolution discussions.

Responsibilities of Project Managers in the Preparation of Schedules

The various PMs comprising the team have a variety of responsibilities in the preparation and use of schedules. These include:

1. **Develop a workable and realistic plan:** PMs should work with all stakeholders to develop the project plan. The schedule will have little value as a management tool without a realistic plan that all can identify with. This is a key element in developing the project schedule.
2. **Develop a workable and realistic project schedule:** PMs should not only rely on their own experience in this process, but also make use of the firm's database, the experience of other PMs, perform a careful analysis of the project requirements, and use any other resources necessary.
3. **Distribute the initial and revised project schedules to all team members:** PMs must recognize that the team extends far beyond their own staff and includes designers, engineers, the general contractor, subcontractors, the owner/client, vendors, suppliers, code officials, and all other project stakeholders—all need to see the project schedules.
4. **Regularly monitor and update the project schedule:** As conditions and requirements change, updating is essential. The schedule is a living document and is only valuable if it is maintained and distributed. On most projects, the scope changes over time and requires prompt and accurate updates to the project plan. Other team members may be significantly impacted by these scope changes and it is important to share the revised project plan/schedule with them.

230 *Project Management for Designers and Facilities Managers*

5. **Project managers must retain cost and time data for use in planning future projects:** The project schedule provides important information useful in planning subsequent jobs. Loss of this historical material could be detrimental to the firm and to other PMs.

Figure 12.6 provides a graphic summary of the planning and scheduling process discussed in this section. While this is a simplified illustration, it highlights the key elements of concern to PMs. It is an interactive and continuous process as the project moves toward completion and requires regular attention on the part of project managers if it is to be of great use to all members of the team.

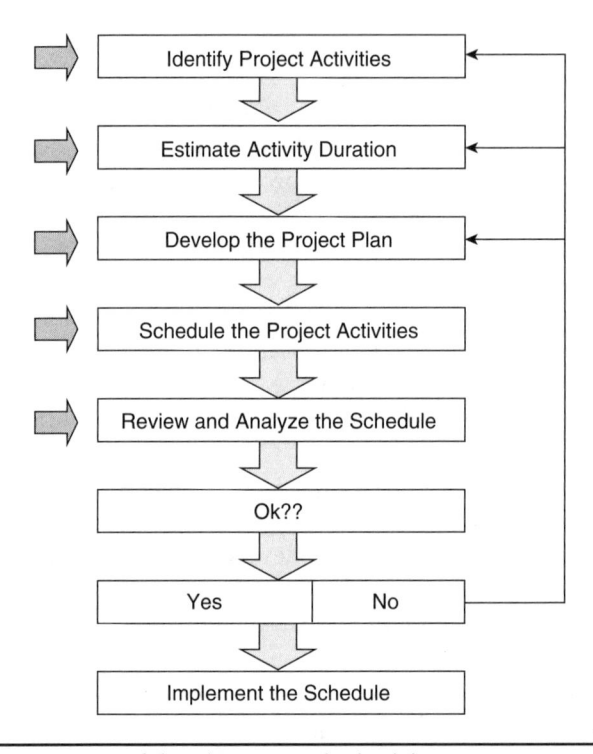

```
Identify Project Activities
          ↓
Estimate Activity Duration
          ↓
Develop the Project Plan
          ↓
Schedule the Project Activities
          ↓
Review and Analyze the Schedule
          ↓
         Ok??
          ↓
   Yes    |    No
          ↓
Implement the Schedule
```

Figure 12.6 Summary of the planning and scheduling process

This book has free material available for download from the Web Added Value™ resource center at *www.jrosspub.com*

13

COMPUTER APPLICATIONS

INTRODUCTION

A Brief Journey through Time

At a recent workshop for engineers, I was asked my thoughts on the next technological revolution in the construction industry. The question prompted me to reflect back on the technological changes that have occurred in the construction industry since World War II.

The 1950s

The two major technological changes to affect construction during this decade were the development of mainframe computers and specialized software for scheduling construction activity. During the decade, the technology of computers developed beyond the early relays, switches, and tubes into more reliable, lower cost machines using transistors. Few firms possessed the resources to purchase a computer, but early service bureaus provided limited access for solving the most complex problems. The big advance was the development of the critical path method (CPM) and the performance evaluation and review technique (PERT). CPM was developed by the DuPont Company in 1956 for use on their complex construction projects. About the same time, the U.S. Navy developed PERT. The primary difference between the two is that CPM uses one time estimate and PERT uses three (most likely, most optimistic, and most pessimistic). Actually, it was the development of a disciplined method of planning sequences that was important, not simply the use of the computer. Both CPM and PERT can be done manually. Today nearly everyone relies on readily available software programs and most people in the construction industry have absolutely no idea how to do these activities manually.

The 1960s

This decade saw computer use in the construction industry take off. Mainframe computers, particularly those produced by IBM (the IBM 360 Model) and Sperry were growing in use and importance in many larger design and construction firms. The major initial application for this computing power was for structural, mechanical, and electrical calculations that could be done more quickly and accurately than using a slide rule. (I still have two slide rules and my college-age son is mystified as to what they are for.) Some firms began to write proprietary software for financial management, but as of yet no software existed for project job cost accounting for project managers' use.

Interest in the computer was growing in the industry and steps taken during the 1960s would have great influence on all of us in the future. On December 5, 1964, an all-day conference on *Architecture and the Computer* was held by the Boston Architectural Center. Nearly 500 architects attended to learn what the electronic future might hold. At the conference, Boston structural engineer William LeMessurier (engineer of the Citicorp Center in NYC, among many others) noted: ". . . About a year ago at a conference, similar to this one, sponsored by M.I.T, the community of practicing engineers first learned of a new approach to the use of the computer. Imagine that you were told that the computer had been trained like a graduate student in civil engineering, not how to solve special individual problems, but trained in the general classical methods of solving indeterminate structures. Furthermore this graduate student could be talked to in simple English and stood ready to go to work on any problem. All that he needed was a geometrical description of the problem with some verbal request for the analytical information required. Wouldn't you think that the millennium had arrived? Well, in a way, it had. This 'graduate student' is a genie called STRESS, an acronym which stands for Structural Engineering Systems Solver; and is well named. At that meeting a year ago, everyone present was able in one afternoon to learn the STRESS language and use the computer programmed with STRESS to solve a problem of his own choice. Program writing was no longer a necessary step to solve each new problem. In retrospect this occasion appears to have been a real breakthrough and its effect on our own practice has been extraordinary. A whole area of engineering activity has been permanently changed..."

Another conference speaker noted: "Putting a whole computer at the disposal of one man for an indefinite period of time would obviously be much too expensive." A report in *ARCHITECTURAL RECORD* on the conference summarized: "It was possible to envisage that not too far in the future architects would be able to receive engineering data and evaluation of functional characteristics almost instantly, at any stage in the design process; and specifications and working drawings of the finished product could be produced with great rapidity using computerized technology."

The 1970s

By the early 1970s, great leaps had been made in both hardware and software. IBM came out with its System/370 Model 145 computer that used semiconductor technology. Wang developed a word processing system that had limited capability, but was widely used particularly by attorneys. The storage capability of this system was tiny and used *floppy* disks the size of old long playing records. Over the course of the decade we would see the early beginnings of the PC/Apple revolution that began with home hobby kits like the January 1975 introduction of the Altair 8000, priced at $439 and using the Intel 8080 processor. This was followed by the Commodore PET 2001 in January 1977 selling for $600 and the Radio Shack TRS-80 also selling for $600 in August 1977. Apple Computer was formed on April 1, 1976 and introduced the Apple II in April 1977, priced at $1300. The decade saw the rapid development of ever more sophisticated and faster chips and storage devices.

On the software side, advances were occurring in both job cost accounting software and in the development of computer-aided design and drafting (CADD). The first significant steps in the development of integrated job cost accounting/financial software for engineers and architects were begun in 1968. Dr. G. Neil Harper, working with the backing of the American Institute of Architects (AIA) and the National Society of Professional Engineers (NSPE), started designing and programming software. These efforts took time and a great deal of money, but by the mid-1970s a workable package was available. Called the Computerized Financial Management System (CFMS) it was the antecedent of many other software packages to follow. The firm Dr. Harper founded is now part of Deltek Software.

CADD was the real revolution of the '70s. The aircraft industry had been developing CADD capability for years, but it was only in the early 1970s that the first real application to the construction industry was started. The major early pioneer in this effort was the Boston-based architectural firm of Perry, Dean and Stewart. In an effort led by Kaiman Lee, the firm invested heavily in the development of the ARK/TWO system. As Lee wrote in July 1971, "It took five years to set up and another year to shake down, but (we) have devised an in-house computer system and interrelated programs that multiply (our) approaches to design." Skidmore, Owings and Merrill (SOM) in Chicago also invested heavily in developing CADD capability. By the end of the decade, CADD in the construction industry was on its way.

My own experience in the early use of CADD began in 1971 with the very first course offered at my architectural school in the use of this tool. As no one on the faculty had the necessary knowledge, it was taught by one of my fellow students. We spent weeks writing code in an effort to draft a line from point A to point B. Eventually, we succeeded, however not in a straight line!

The 1980s

The PC revolution soared with the August 1981 introduction of the IBM 5150 PC computer. These early computers had limited capability and even less storage. But, they were a major breakthrough in bringing computers to the masses. Followed by the XT, AT, and 386 models, IBM brought capability up and prices down. It seemed that everyone from Adam Osborne to Timex began to manufacture personal computers, but over time, few companies would survive the price cutting and weekly obsolescence of their product line. Apple Computer thrived by focusing on the advanced graphics capabilities (GC) of their machines. IBM products were inferior to Apple products, but IBM licensed its technology to other manufacturers beginning with Columbia Data Product's MPC in June, 1982 (Apple did not license its technology) and stepped up its own marketing. In time, PC-based machines with Intel chips would dominate the worldwide market.

CADD became ubiquitous as software platforms stabilized under a couple of manufacturers and the federal government started requiring drawings be completed on CADD for its projects. Prices for CADD software declined dramatically. Initially, there was great turmoil in the CADD marketplace. Many diverse manufacturers entered the fray. For example, in 1983, Bausch & Lomb (the optics firm) joined the heated competition by unveiling ProDraft from its Interactive Graphics division. Priced at $29,995, it was aimed at smaller architectural and engineering firms, those employing 15 or fewer designers. Later in this chapter, Gene Montgomery discusses how CADD software works.

Communication technology also changed during the 1980s. The mid-decade period saw the availability grow and the price drop for fax machines. This simple device allowed instantaneous transmittal of text, photographs, drawings, and other material from any location with a telephone line. Cell phones became available and rapidly became easier to carry. My first cell phone came in a shoulder pouch carrying bag that was portable—just. Service was spotty and expensive, but was a precursor of things to come.

Off-the-shelf computer software for scheduling, accounting, spreadsheets, engineering, word processing, and many other applications became cheap and very available. It was during this time that the game of continually coming out with slightly changed, new (but often not improved) versions of software every six months was perfected by Microsoft and others.

Firms providing job cost accounting software for the industry multiplied during the decade. Visits to the annual A/E/C SYSTEMS shows found many new hardware and software providers. Often, these providers were not in business by the time the next show came around.

The 1990s

This being more recent history, many readers will be familiar with the changes this decade brought. PCs continued to grow in power, speed, capabilities, and use during the '90s. Software was ever cheaper and more diverse. The development of wireless technologies allowed contractors to remotely locate their equipment, monitor its use, and track costs and charges (and bill for them accordingly). Today, wireless technology continues to expand in applications and benefits to the industry.

Cellular technology expanded and nearly everyone in the construction industry had a cell phone and could be in almost continual touch with their team. New technologies such as Personal Digital Assistants (PDAs) provided another tool to access and receive information.

The most important development of the 1990s was the growing availability of the Internet. Mosaic, the first web browser was introduced in 1991. I received my first version of Mosaic in 1993. As a novice, I found it daunting, but as I learned more, I began to appreciate the potential of the Internet. 1994 was the first big Internet year. Since then there has been a never-ending stream of ideas and applications to make use of this wonderful tool (the April 2000 dot.com/NASDAQ collapse notwithstanding).

For the construction industry, the other big news from the 1990s was the development of web-based project management software. Some trace the origin of this software to 1995 when e-Builder introduced project website software. Many others point to 1997 when large firms such as Lend Lease Corporation (Sydney, Australia) wrote their own programs to allow collaboration over the Internet. Also in 1997, software was written to allow project collaboration on the new Charles Schwab headquarters building in San Francisco. The first significant commercial service grew from this latter effort and became Bidcom.

At the spring 1998 A/E/C SYSTEMS show there were no exhibitors offering web-based project management software. By the spring of 1999, there were several. By the spring of 2000, nearly the entire show was devoted to Internet related products and services. Soon after, there were many mergers and failures of web-based collaboration software providers. By 2004, the Center for Design Informatics at Harvard University estimated that there were still as many as 175 firms offering some form of web-based project collaboration products to the construction industry. The Center was originally established to study the impact of web-collaboration on the industry and tracks available products.

The 2000s/2010s

Despite vendor claims to the contrary, web-based collaboration in the construction industry has never lived up to its potential. While some industry leaders are still pushing the envelope, most firm's adoption is limited by a lack of participation by all members of a construction project, nagging concerns (real or imagined) about the security of data, and the lack of interoperability between software platforms. It now seems clear that the use of the Internet for project collaboration and management will be evolutionary rather than revolutionary. Some industry vendors such as Primavera and Autodesk are steadily broadening their product lines to incorporate more of the capabilities sought by the construction industry and *enterprise* software is now widely touted. Later in this chapter, Robert Schneider provides an in-depth look at using the web for project management.

The mid-to-late years of the decade saw the development of Building Information Modeling (BIM). According to an entry in Wikipedia, "BIM covers geometry, spatial relationships, geographic information, quantities, and properties of building components. BIM can be used to demonstrate the entire building life cycle including the processes of construction and facility operation. Quantities and shared properties of materials can easily be extracted. Scopes of work can be isolated and defined. Systems, assemblies, and sequences are able to be shown in a relative scale with the entire facility or group of facilities." As happened with CADD, BIM is now being required by many federal government agencies. Unfortunately, no single software developer's platform is the standard, causing challenges for firms in the industry using (or required to use) Building Information Modeling.

Also, the middle years of the 2000s saw the widespread use of Smartphones, originally led by RIM's Blackberry phones. By 2009, Blackberry was dominant, only to quickly lose its lead to Apple and others. Tablets were widely adopted after their introduction by Apple in 2010 and soon other manufacturers entered the fray. (Although various basic versions of tablet computers have existed for years, it was Apple's marketing which popularized the product category.) Laptop computers became thinner, lighter, and more powerful. By 2015, improved products were still entering the marketplace making the work of project managers (PM) in the construction industry easier. However, the introduction of new innovations and product categories have slowed. Software continues to evolve and improve, but much work remains, particularly in the operation of BIM products. (More on this later in this chapter.)

Some observers of the computer industry now perceive that it has reached the "mature" stage of development. Microsoft's introduction of new or improved operating systems (think Windows 8!) has become largely irrelevant and Intel's ability to produce faster and cheaper chips has become more challenging. Even Apple, long a computer innovator has become a consumer products company producing the iPad and iPhone. Some real advances continue in wireless technology and these are readily being adopted by the construction industry.

WHAT DO CADD SYSTEMS DO?*

General Comments

CADD equipment can do some tasks incredibly fast and with great precision. A drawing which took days to draw can be copied in seconds. A

---

*This material was prepared by Gene Montgomery, AIA.

portion of a drawing can be mirrored or rotated or scaled differently and placed in another location almost instantaneously. These are powerful capabilities. It takes experience and forethought to know how to organize the drawings in order to use these capabilities efficiently.

With the computer, you are not restrained by the size of the paper; composition can be completely rearranged in seconds; details can be revised beyond recognition with minor rearrangement. These qualities produce work rapidly—but they can also multiply mistakes. Good judgment, planning, and understanding of the results of actions are essential qualities of a computer graphics operator.

Computer graphics files do not contain lines and arcs. They are collections of numbers which locate the various entities in a coordinate system. Every entity has a coordinate associated with it and the file always *knows* exactly how far one entity is from another. All systems use this quality to produce semi-automatic dimensioning. With the computer, there is no such thing as a rough sketch. All entities are located very precisely in space. With this capability, machine drawings begin to diverge from manual drawings. A line doesn't exist as a line on a piece of paper; it is a line that connects two unique points in space. The line *knows* where it is and can tell you exactly where it is any time you ask. The simplest computer graphics system has files that mimic manual drawings but also contain information not existing in manually prepared drawings.

More sophisticated computer graphic systems add information to the entity. A line can be told not only its exact position in space, but also what it is or represents. For example, a line may represent the centerline of a wall. Various types of information can be encoded into the drawing and, with the proper editing procedures, reported to the user in predetermined formats.

Not only can entities in a computer know what they represent, but they can also know they are members of a specific group. And the group may know information about itself independent of the sum of its parts. Drawings can become repositories of great amounts of information not directly related to the graphics. This file information can be edited and manipulated by drawing editing techniques which may be more easily understood than the numeric manipulation of conventional computer files.

Properties of the file can be amplified and supplemented by software and peripheral devices that use the computer files. The file can be *read* and viewed on a screen. When the drawing is plotted by a printer, it will look much like a traditional, manually produced drawing. If the computer file was prepared as a 3-dimensional drawing, software can transform it to a perspective drawing. More sophisticated software can read the drawing, analyze the information associated with the entities and reach conclusions based on that information. For example, information describing the construction of the exterior walls of the building, the windows, and their orientation can be encoded into the drawing. Software can then read this information and calculate the heat-loss for the building.

Further along the path of sophistication is software that can change the graphics based on embedded characteristics. For example, the computer can be told that all 3' wide doors are now 3'4" wide and the software would change the graphics accordingly. Despite the capabilities of even the simplest CADD system, some users fail to get past the step of using the system as a substitute for manual drafting. The reasons for this are many:

1. **All manufacturers strive to make their products *user friendly*:** This means that they are easy to learn and use, and are tolerant of mistakes made by the user. There is a basic contradiction between tolerance of errors and the computer's innate precision. Computer language is an elaborate code with a very precise syntax. Each time error tolerance is built into the code, it becomes more complex. If the machine makes judgments about the meaning of the data entered, then potential errors are built into the data. If the machine asks for clarification, then additional time is required for data entry. The contradiction is usually resolved by requiring accurate data entry.
2. **Graphic data is comparatively easy to check visually:** If Line A is too far from Line B, it will appear in error and the distance can be measured and verified. But if Line A is encoded to mean that it is the surface of a brick wall that is 8 feet high, then the data entry problem becomes more complex and the additional information is not visually apparent. To use the data for heat loss calculations, you must know where the other surface of the wall is, what the internal construction is, and whether the surface is an interior or an exterior surface. A person sees and understands this construction at a glance. A computer can understand it only if the data is entered completely and accurately. It is possible to *teach* the computer rules about how to resolve ambiguous or incomplete information. But to keep the rules manageable, a great deal of standardization is required.

3. **Software that performs specific tasks, like computing heat loss calculations, has a comparatively small number of potential users:** As a result, the cost of development becomes quite high. In addition, few engineers make their calculations in exactly the same way. A small market, a nonstandard process, and a variety of proprietary computer systems all conspire to make the cost of software to perform a task—which can be done quickly on a manual basis—relatively high. As a result, many users never get beyond using their machines for basic drawing work.

WEB COLLABORATION TOOLS—WILL THEY FIT ON THE A/E/C INDUSTRY TOOL BELT?*

Introduction

Two rather low-key events took place during America's transition from the Industrial Age to the Information Age. The first personal computers were sold in 1982, and the World Wide Web was introduced to the general public in 1994. These seemingly small events illustrate that widespread dissemination of information is the key to creating a *knowledge economy*. A direct outcrop of the Internet phenomenon is the proliferation of web collaboration tools for the A/E/C industry. There are a number of firms vying for market shares in this area. The A/E/C industry has awakened to the impact of web collaboration and is attempting to sort through the hype. Universal acceptance by the industry is not assured, as questions of project delivery systems, pricing, information control, and cost benefits are examined. This material discusses the issues firms need to consider as they begin to incorporate these management tools into their business systems.

The Information Technology Promise

The enduring wisdom of William Shakespeare offers us this pearl: "What is past is prologue." Sorry, Bill—it's a pithy phrase, but the brave new world of technology has piled a lot of established wisdom on the trash heap. Wisdom, outlooks, and work philosophies are changing with the

---

*This material was prepared by Robert S. Schneider, P.E. It originally appeared in the April 2001 issue of *Civil Engineering News* magazine and has been edited here. While some of this information is dated, it provides a very useful discussion on the use of the Internet in design, facilities management, and construction firms.

same velocity of an electron traveling down a fiber-optic cable. The information technology pledge to the A/E/C industry is quite simple: web collaboration tools reduce project delivery times and subsequent costs through improved communication processes. However, given the complexities inherent in the design and construction process, one questions whether these tools really can *improve* the process and save money.

The most common pitch among the web collaboration developers is that of saving document reproduction and shipping costs. Anyone in this industry knows that trees are sacrificed daily so that design and construction offices can leave a paper trail. The new tools promise to diminish the idea of printing to a second alternative since they allow almost any file format to be viewed online with a commonly used web browser and a document viewer. Other features include the following:

- Automatic, external e-mail notification can be sent to team members when relevant documents have been added.
- Notifications can be sent for new task assignments, meeting requests, etc.
- Viewer technology allows for the display of hundreds of different file formats, including CADD documents, without the application being present on the computer.
- Built-in collaboration tools let team members add comments or indicate changes on drawings anytime, thus accelerating the decision-making process.
- Built-in forms for processes, such as submittals and daily logs, allow team members to track and respond to situations online.
- Audit trails are created with logs developed automatically on every file. The unprecedented access and centralized storage abilities of these systems induce project responsibility. In other words, you can run, but you can't hide.
- Security features in the internal and external environment allow for selective viewing by team members. The web-enabled tools also help organize the structure of the project team. This is not to say that these tools will flatten organizational structures, but they can make communication more direct, allowing the team to be more efficient. In short, it can easily put information into the right hands, at the right place, and at the right time—with point-and-click accuracy.

Flexible Project Delivery Systems

If you visit any web collaboration site, it will basically give you the same pitch regarding an improved communication process. After all, that is what they are selling—a process to bring your projects in on time and within budget. Most companies offer only one version of their web collaboration product. As it stands, this product must meet the needs of owners, designers, and contractors. As time goes on, we may see niche products. The ability of these products to be flexible in the context of multiple project delivery systems and organizations will be the deciding factor of their acceptance.

As no two painters approach a canvas in the same manner, seemingly identical projects are rarely approached in the same way. In fact, the trend in the A/E/C industry is to increase flexibility in the project delivery system. The majority of municipal agencies use the design/bid/build system. However, a discussion paper prepared by the National Society of Professional Engineers stated that the majority of private-sector projects would be built on the basis of a design/build arrangement in the near future. And, since so many firms market their services to both the private and public sectors, they must work with both project delivery systems. However, by attempting to sell just one version of their product, web collaboration developers may not be able to serve the entire A/E/C industry.

Site Administration and Information Control

As far as I know, the utopian vision espoused by the communes of the 1960s, where everyone was free to achieve self-actualization, also needed some kind of organizational structure. Web collaboration tools are no exception; at least one team member must assume leadership. The more sophisticated web collaboration tools require one person to act as a site administrator. This person should be a PM who understands each team member's responsibilities. In addition, this administrator would set security levels for each person and each file on the site. The project delivery system directly influences who handles these responsibilities. Because the roles of the owner, contractor, and architect/engineer (A/E) firm differ with various project delivery types, so do the choices for site administrators. Typically, the contractor acts as prime to the owner, with A/E firms acting as sub-consultants in a design/build arrangement. In this case, the contractor is more likely to contract with a web collaboration provider and act as the site administrator. But in the design/bid/build system, the contractor is not involved in the design process, and it is probable that the engineer would take control. However, the question remains—who assumes responsibility for a project site during the construction phase if the site was provided and administered by the design consultant during the design phase?

Questions about design drawing ownership are also raised in regard to online project management. Some design professionals feel that they own the drawings they produce. By placing their work on another firm's server (*or now on the cloud*), they feel that they lose control of this information. However, given the proper tools, both electronic and paper documents can be altered—regardless of their existence on the web or on someone else's server. To combat this apprehension, I think the web collaboration developers need to do a better job of explaining to the industry the technical and legal implications of transferring project information.

Another difficult item to quantify is improved communication. In the most basic sense, improved communication means the ability to communicate your ideas through text and graphics more efficiently. The installed document viewers do just that. Just about any project documentation from drawings to project schedules can be easily placed and viewed on the project site, thus reducing the need to fax or mail documents. The real benefit of the viewer is that it allows non-CADD users to access drawings at any time without having to ask a CADD user to run a plot. Additionally, it enables PMs from different organizations to discuss a drawing over the phone while simultaneously viewing the document.

Accelerated Decision-making Process

One of the most promising possibilities in reducing project costs by using web collaboration is accelerating decision-making processes by minimizing communication errors. Of course, this is another impact that will be difficult to measure. For example, web collaboration tools have the ability to accelerate project schedules by changing the way designers approach the submittal review process. Simultaneously, the tools meet the needs of the plan reviewer.

As an instructor once said to me, "There is one thing you should always know about project schedules—they are wrong." He wasn't suggesting that anyone abandon the process of developing a work breakdown structure and Gantt chart; he was just pointing out that forces beyond our control influence most project schedules. Anyone who has worked on public works projects that require citizen input knows what I mean. But can these tools help accelerate the process and potentially reduce project costs? I think they can, not because external influences will diminish, but because online collaboration can change the current budget-driven process to an issue-driven process.

In terms of time, one indefinable aspect of projects is the submittal review process. While many municipal agencies have developed submittal-processing guidelines in response to public pressure, many agencies have not. Given the complexities of staffing and the multiple project environments in which most organizations operate, schedule management is largely an individual responsibility.

As you may know, public agencies periodically review work performed by design consultants to check for compliance with agency standards and project-specific requirements. The current, budget-driven process is typically defined by a percentage of the budgeted cost of work completed. In other words, this process requires submittals at 30-percent, 60-percent, 90-percent, and 100-percent stages. In theory, this means the consultant has completed a percentage of the project scope measured in terms of dollars.

Each submittal milestone can be a time-consuming process that includes collecting, plotting, printing, and shipping the construction documents to the public agency. Many design consultants are reluctant to continue working on a project until the review period is complete. This can take anywhere from three to five weeks for each review, depending on the size of the project. This usually means working on another project until the comments are returned. As currently practiced, the submittal process can add anywhere from 10 to 15 weeks to the life of a project. Web collaboration tools could enable projects to flow differently, adopting an issue-driven process of review rather than the budget-driven process described above. A review of some project management theories is necessary to explain how such a system would work.

Classic project management is considered to have the following five phases: planning, organizing, staffing, controlling, and directing. Each phase, like each project, is full of decisions—some are important; some are not so important. So which issues are important? The 80/20 Principle, developed by Italian economist Vilfredo Pareto, can help answer this question. Pareto observed that 20 percent of the decisions produce 80 percent of the results. Or stated another way, 20 percent of the (Pareto) issues are critical to achieving the project objectives. If those issues can be identified and addressed, there is an 80-percent chance that the project will achieve its objectives. So when do Pareto issues occur in the life of a project? Successful design consultants can identify most of the Pareto issues during the *go/no go* decision process, while the remaining can be identified during proposal preparation. Therefore, in the budget-driven process, submittals at 60-percent and 90-percent levels have little impact on a project's success. If issues arise that substantially alter the scope of work or if the objectives change, the client likely changed his or her mind, or the project was planned poorly.

For an issue-driven submittal process, the interaction between the design team, the owner, and other stakeholders is defined by a series of mutually agreed to Pareto issues and decisions. Web collaboration tools enable project teams to communicate issues *continuously* by providing constant access to all team members. In a slightly delayed, simultaneous fashion, a continuous review process based on Pareto issues can replace the prevalent budget-driven submittal process. This can reduce communication downtime significantly.

Web Collaboration Potential

Web collaboration has the potential to change the way the industry approaches information flow, which may ultimately reduce communication errors. In a knowledge-based economy, our actions can be reduced to two simple processes: the development of information and the ability to communicate that information. Web collaboration is all about efficiency in communicating information. The design and construction process for the last 50 years has been predictably balanced by complying with a statutory induced, budget-driven process. The new, web-enabled tools have the potential to accelerate project schedules by creating imbalances within an issue-driven process.

Embracing these new tools and switching processes will, in the short-run, create a struggle between digital visionaries and the digitally-challenged. Eventually, the variety of products will expand, costs will be absorbed, and information control and transfer will be routine. Once this technology is embraced by the A/E/C industry, even the construction laborer will become a knowledge-based worker.

BUILDING INFORMATION MODELING: PROMISE UNMET?

Historical Background

Fifteen years ago, web-based project management in the construction industry was the hot new technology. The annual *A/E/C Systems* show was overrun with products, promotions, and services using web-based software. Vast amounts of venture capital were poured into software development and product marketing. Web-based project management was to revolutionize how the construction industry worked. Owners/clients were hopeful that new relationships among team members could be forged, coordination improved, options multiplied, and construction costs contained. Harvard University established an institute called the Center for Design Informatics to study the impact of this technology on the industry. *Engineering News-Record* (*ENR*) and many other publications ran numerous articles on the subject.

Unfortunately, web-based project management never quite worked out. The reasons were many (more on this later). Fast forward fifteen years and we are now being inundated with glowing reports of BIM. It's the new Holy Grail, promising everything and much more than web-based project management was to provide. As with web-based project management, *ENR* is again filling its pages with articles about this latest technology. Seeing BIM in action is impressive, but will it lead the construction industry to a new promised land of profit, cooperation, and quality? Disciples say yes—others (including me) have seen this all before and have doubts. What are some of the concerns related to BIM?

1. **Lack of interoperability of software products:** There are two major vendors of BIM software—Autodesk, Inc. and Bentley Systems, Inc. In 2008, they announced an agreement to exchange software libraries with the intention of improving the ability to read and write their respective formats. While these two companies dominate, many other vendors also offer BIM software. Multiple incompatible platforms mitigate a major benefit promoted for BIM. As *ENR* notes (March 2, 2009, p 30): "Immediate gratification on a project, which can have a dozen BIMs, doesn't exist when there are different disciplines and various trades using myriad applications." *ENR* also commented: "For the time being, there is lots of BIM babble." Gordon Holness, president-elect (2009) of the American Society of Heating, Refrigerating and Air-Conditioning Engineers (ASHRAE) commented regarding interoperability: "We do not seem to have made much progress in the last few years."

   While the U.S. General Services Administration and other owners are supporting specific platforms, many other products are widely used. Some critics also point out the lack of scalability of the software and problems with remote collaboration as continuing issues. These were all problems with web-based project management as well. They were never resolved.

2. **Lack of use by all team members:** The construction industry is made up of hundreds of thousands of small businesses, representing hundreds of disciplines, specializations, and segments. BIM, like web-based project management before it, works best/successfully when all team members use it (and typically the same software). Gordon Holness of ASHRAE has noted: "At this point, less than 20 percent of the mechanical-electrical-plumbing engineer (MEPE) projects are modeled, and that number is going down, not up" (*ENR*, March 2, 2009, p 32). Part of the cause for low adoption of BIM software was identified by Greg Lakota, of Halvorson and Partners, Chicago, "Software capabilities often fall short of what is promised to and expected by the user."

Atul Khanzode, director of virtual building for DPR Construction in Redwood City, CA told *ENR*, "The majority of the industry, including vendors, is currently in a hype cycle when it comes to BIM, and as we like to say, everyone seems to be jumping on the BIM-wagon." He continued, "Technology vendors are taking this opportunity to sell software licenses. Very few are taking the time to create best practices for the real-world application of the technology to achieve the kind of results the industry is starting to expect." As is often the case, vendors are overselling the capabilities of their software. This creates expectations in the mind of owners. Consultants adopt BIM in an effort to meet these expectations and nearly all are disappointed. This creates a feedback loop where BIM use will decline until/unless software issues are resolved. If the problems are not solved, BIM will go the way of web-based project management.

3. **Concerns about the control, accuracy, ownership, and sharing of data:** This same issue appeared regularly with web-based project management. Who controls the information? Who ensures its accuracy and coordination among design disciplines? Architects would certainly like to assume this role; however many are ill equipped for the task. The AIA is a strong proponent of BIM and has developed a series of new construction documents to promote Integrated Project Delivery (IPD). IPD shares project risk and reward among the various members of the project team including the owner. Architects assume a central role when IPD is used. Is

the AIA using this as a device to preserve/enhance an architect's market share? For IPD to work effectively, sharing of information is vital. Standardization of electronic files is an essential part of this approach. IPD and BIM therefore have many of the same requirements. Skeptics could ask, "Is BIM partially a marketing device for some designers or a true benefit to the owner?"
4. **Legal concerns:** Along with the new BIM contract documents come unresolved concerns. In the event of an error or omission, who is the responsible party? Who handles coordination? Who is responsible for seeing that information and updates are communicated to all team members? Until a body of court rulings appear, the construction industry is in unknown territory.
5. **Inefficient software:** Engineers in particular have expressed concerns about everything from poor graphics, to inefficient internal structure, to ineffective interfaces with their specialized programs regularly used for structural, mechanical, energy, and other calculations. *ENR* reports, "Design packages used down the line—from analysis software to a BIM to the steel fabricator's application, to the mechanical-electrical engineer's application—can all be different. Designers say they waste a great deal of time, due to communications breakdowns between different packages and even between different applications within a single package." On the other hand, BIM has proven to be very effective at *clash detection and visualization*. It does save rework in the field and is popular with contractors. Significant productivity gains have been reported.

Software problems with BIM remain significant. For example, an article on CENews.com (January 2015) called *Seismic Performance Analysis and BIM*, notes: "In its current state, BIM software does not have the capability to account for so many different methods of fortification analysis without finessing a model through many pieces of software or custom programming, which can be a cumbersome task. Design models used for construction documents must be passed from a BIM model to an analysis model. Even when this is done through a single company's software suite, disconnects occur between BIM and analysis models while round-tripping results. When attempting to use third-party analysis software, this disconnect can be even greater." The article continues, "With new versions of software and better integration of analytical applications, BIM will eventually provide engineers with all the tools necessary to construct building-specific data that is needed." Perhaps the key word in this statement is—eventually.

ENR concluded a recent article by forecasting, "The (BIM) technology monsters will likely be tamed by 2020." Can the construction industry wait? BIM does have potential, just as web-based project management had potential. Time will tell if BIM will be the game changer its predecessor was not.

CASE STUDY: BUILDING INFORMATION MODELING (BIM)

From: ENGINEERING NEWS-RECORD (ENR), MAY 23, 2011, Nadine Post, author

"A Cautionary Digital Tale"

(Reprinted courtesy of *Engineering News-Record*, Copyright Dodge Data & Analytics, 2011, all rights reserved)

A lawsuit over construction of a life-sciences building at a major university stands as the first known claim related to the use of Building Information Modeling by an architect. Furthermore, the claim and its settlement serve as a cautionary tale to others using BIM, says the insurer.

"The creators of BIM claim its use reduces risk, and indeed it can—like any other tool, if it is used right," says Randy Lewis, vice president of loss prevention and client education at the Denver office of XL Insurance, which provides professional liability insurance to licensed design professionals. "If you don't use BIM correctly, you can get into trouble."

For the life-sciences building, the architect and its mechanical-electrical-plumbing (MEP) engineer used BIM to fit the building's MEP systems into the ceiling plenum. But the design team did not tell the contractor that the extremely tight fit, coordinated in the BIM, depended on a very specific installation sequence.

When the contractor was about 70% through assembly, it ran out of space in the plenum. "Everything fit in the model, but not in reality," says Lewis. The contractor sued the owner, the owner sued the architect, and XL brought in the MEP engineer. "It was a very costly claim to negotiate," says Lewis. XL did not litigate the claim because it would be difficult for any jury to comprehend."

Lewis declines to offer specifics on the project, other than to say the building is open. He also declines to name the players. As far as the settlement goes, he will only say there was a "pretty significant cost," totaling millions of dollars, which was shared by the architect, the MEP engineer and the contractor.

"The problem was poor communication. The design team never discussed the installation sequence with the contractor, and the contractor wasn't sophisticated enough to understand the importance of assembling the components in a certain order," says Lewis.

Insurers advise designers not to get involved in means, methods, safety and sequencing. But, for this project, even though it was delivered under a traditional design-bid-build contract, it was not enough for the architect to say, "I designed it, it fits, here you go contractor, figure it out," says Lewis. "In the BIM world, parties are encouraged to communicate with each other and make suggestions." That did not happen on this project, he adds.

From *Engineering News-Record (ENR)*, June 6, 2011, "Cautionary Tale Triggers Heated Responses About Digital Modeling." (The following are various letters/posts received by ENR in response to the article and included in the June 6 issue of the magazine.):

"It sounded as though the contractor didn't participate in the BIM process. If the model was fully coordinated and was supplied to the contractor for reference, the contractor could have used 4D sequencing tools to examine the various work flows and discover this issue in the virtual realm. If the virtual model fits and 'works,' then it's the contractor's responsibility to reproduce the bids."

"Lack of communication! That says it all. An hour with a tape measure and a set of prints prior to MEP installation would have avoided that nightmare!"

"It's a shame that even though the design team did enough due diligence to figure out that a certain order of construction was required, they never told the contractor about it."

"Thank you, *ENR*, for the caution. Simply having a BIM that shows that everything fits may not be enough. It is up to designers and contractors to use this information to inform how they want to proceed on their own projects."

APPENDIX 1

PROJECT CLOSEOUT

INTRODUCTION

General Comments

After the punch lists have been completed and final actions taken, the design firm project manager (PM) remains an important part of ensuring client satisfaction. Regular follow-up with a telephone call or meeting will let the client know you are concerned about service and his or her satisfaction. Some firms request client completion of a report card form (see Figure A1.1). Time and expense spent giving attention to problems should be immediate and may not be billable, but may need to be charged to marketing. The client should be kept informed of your firm's activities and should receive all relevant mailers. Some design firms enter into a formal Commissioning process to assist their clients in the operation and maintenance of sophisticated mechanical systems. It is important to follow up to ensure client satisfaction, learn from the experience, and obtain additional work if available.

Project Data Retention

Once a project has been completed, a decision must be made as to the fate of the vast amount of data and material collected during the project's life. No firm can or should save everything. Many of the materials that accumulate are redundant, are working or draft versions, or are progress reports. The pack rat firm that saves everything will soon be overwhelmed.

Date: _____   **Return to:**
                                 Bill Smith
To: _____   ABC Design
                                 123 Fourth Street
Re: _____   Albany, NY 11111
                                 Telephone (518) 666-6666
Job #: _____   Fax (518) 777-8888

Project Manager: _____

On a scale of 1 through 5, (1 being the lowest mark and 5 being the highest), how would you rate ABC on the following?
    1. Listening to your needs and understanding the project.   1 2 3 4 5
    2. Technical capability.   1 2 3 4 5
    3. Innovation.   1 2 3 4 5
    4. Quality of our work.   1 2 3 4 5
    5. Our responsiveness.   1 2 3 4 5
    6. Frequent follow up on questions and issues.   1 2 3 4 5
    7. Meeting your schedule(s).   1 2 3 4 5
    8. Meeting your budget(s).   1 2 3 4 5
    9. Quality of our project management.   1 2 3 4 5
    10. Quality of our staff.   1 2 3 4 5

Compared to prior years, or projects, have our services improved.   Yes   No   (circle one)

Comments: (If more space is needed, please use the back of this form.)
_____
_____
_____

Please answer the following:
    1. Were you or your client ever disappointed over something we did or did not do?
        Yes   No   (circle one)   if yes, please explain.

    2. Would you like to have a post completion meeting on this project?
        Yes   No   (circle one)

    3. If we could do something better, or provide an additional service that would help you, what would it be. Please list as many as needed.
    _____
    _____
    _____

    4. Will you consider ABC for future work?   Yes   No   (circle one)

    5. May we use you as a reference?   Yes   No   (circle one.)

    6. Other comment or perception of ABC:
    _____
    _____

If you feel ABC did an excellent job, we would welcome a "letter of recommendation" that we could use for our business development efforts. Thank you.

Client: _____   By: _____

Figure A1.1   ABC Design report card form

A firm that discards nearly everything will soon find itself regretting the decision. Equally at risk is a firm that retains the wrong materials or one that fails to organize them in a useful manner. Increasingly, firms are saving much of the needed data electronically. Not only does this save space, but also many clients are now requiring their own copy of this electronically stored information.

Some materials must be retained. As-built drawings should be kept as long as the building stands. Your firm may be the only source of these important documents. Future engineers, architects, and building owners will be saved much time and expense if they have these drawings available. The project specifications are also vital to enable others to review the material and equipment used. For historically significant buildings, additional information such as preliminary design drafts may be retained. This will allow future historians and preservationists to consult documents that indicate the designer's thought processes. However, the determination of historically significant is for the future to decide, not the ego of the designer. One clue though, is the original purpose of the building. Single purpose buildings, especially for a branch of government or for a single corporate user, often have the greatest initial impact and the longest life span.

Legal Concerns

Usually, the single overriding factor in determining the retention of data is concern over the possibility of future lawsuits. Most states have statutes of limitations for the filing of legal actions in construction projects. In general, these statutes will dictate how long you should retain certain information. In most cases, 10 years will be adequate. Check the statutes in the states where you work. Typically, materials retained because of legal concerns include a final set of drawings at the completion of each project phase. The owner-designer contract and amendments (e.g., change orders) must be retained permanently. Correspondence that could be used to pinpoint responsibility or a standard of care must be saved. Many attorneys would advise keeping all correspondence until the statute of limitations expires. In addition, job notes, diaries, and field inspection reports must be retained.

Management and Marketing Use

Aside from the legal and historic uses, the most important reason for reviewing and organizing project material lies in its subsequent management and marketing applications. Many engineers and architects are contacted by owners years after the original project completion. This can be the source of much additional billable work and requires ready access to past project data. Obviously, it is essential to save and organize critical information to allow for a timely response to client needs.

Project Managers and Marketers

In many design firms the PMs have the most complete and accurate files. Often, essential data never make their way to the main files of the firm. PMs may leave the company, taking irreplaceable records with them. Even when they remain with the same firm, the information they have is rarely available to other PMs or to the marketing staff. If the main files have gaps, then much time is wasted seeking material, records, and reports that should be readily available.

Your marketing staff has a continual need for historic data on projects. Not only must the appropriate information be retained, but it must be organized to allow for quick retrieval. Without this organization, a great deal of potentially productive time may be wasted searching for information on completed projects. For example, many proposals require submission of information on past similar projects. This often includes fees, consultants used, owners' names and addresses, contractors and subcontractors involved, and staff and PMs' names. Some of this may be easily recalled on significant recent projects; however, this is often not the case.

In an effort to organize and ensure comprehensive records, some firms have established manually or electronically completed project files. These provide a checklist of required information and centralize all essential historical material. They are not a replacement for the firm's main files. The completed project file serves as a supplement, containing key data of value to other PMs and marketers.

APPENDIX 2

PROJECT ENVIRONMENTAL CONSIDERATIONS

INTRODUCTION

General Comments

Project managers (PMs) do not need to be environmental and planning experts. However, they must be familiar with the issues and concepts related to these important topics. Increasingly, environmental constraints impact the design, construction, and operation of buildings, bridges, highways, and many other facilities. Not only are these professional concerns, but they are clearly societal issues. Energy costs, water availability, waste removal and management, traffic, urban sprawl, and many quality of life issues strongly impact the success or failure of an individual building or facility and can directly impact the performance of an entire business.

In many areas of the U.S., recognition is growing that our present wasteful design and construction process is not sustainable. Persistent drought in the West is forcing a reassessment of planning and landscape design that wastes vast amounts of fresh water. Sprawl destroys enormous areas of the landscape, encourages dependence on energy-wasting vehicles, creates air pollution and forces a large percentage of the workforce to spend a great deal of valuable, nonproductive time sitting in traffic. Every design and owner PM is required to have an awareness of the consequences of ignoring these problems. Most important, they must have an understanding of ideas to remediate the problems created by environmental issues and suburban sprawl.

This appendix highlights some of the innovative methods being developed to create a more environmentally friendly design and construction industry. Designers must encourage their clients to consider some of these methods. Owners should recognize that it is in their best interest to examine these options. They can create better-performing buildings or facilities at the same or lower cost than traditional buildings particularly when operating costs are considered. Everyone must recognize that energy costs will continue to rise, water shortages will grow more acute, land availability will diminish, and the cost of labor and building materials will increase. Public officials need to be educated or elected who are aware of the consequences of their decision making. Both pro-growth and pro-environment extremism must be avoided. Sensible planning and design are crucial and tax incentives can play an important role in encouraging environmentally and economically sound construction.

In many urban regions of the U.S., residential developers have found that developable land is in very short supply. Some are turning back to formerly abandoned or underutilized property in older cities or suburbs. Often contaminated, these *Brownfield* properties require special handling and designers need to develop the expertise to aid their clients in this effort. Retailers as well are following the marketplace and moving into shopping districts in formerly blighted areas. Unfortunately, many leaders of the nation's cities and metropolitan areas have not yet realized that endless suburban sprawl is bad public policy, creates a lower quality of life for residents, and is unsustainable. There are, however, many other leaders who have recognized that green planning and design is good for their communities, their residents and businesses, and for their local and regional economies.

THE GREEN BUILDING REVOLUTION

Green buildings (also called sustainable buildings) are more resource efficient than regular buildings simply built to code. Many advocates of green buildings also believe that they improve their occupant's productivity, health, and comfort. One green building supporter notes that directly or indirectly, buildings consume about 70 percent of the nation's electricity. Design and materials that could reduce this consumption would not only save energy, but also reduce CO_2 emissions. The construction industry also uses vast amounts of timber, water, minerals, and other materials in various forms and produces huge quantities of waste materials.

Electrical Generation

A *green* building is not created simply through the materials used in its construction—many elements incorporated into the design qualify a structure as *green*. For example, the use of natural lighting saves on electrical use, maintenance of fixtures, and may improve the comfort of the building's users. Solar panels can be used to generate electricity for internal use and the excess can be applied against power purchased from a local utility. For example, some states such as California make it mandatory for the local utility companies to have bi-directional power meters where the excess (up to a limit) generated by a user goes back into the grid. The user is not *selling* power, but gains a credit against their own usage. The operational savings can be substantial although system maintenance costs must also be factored into the cost/benefit evaluation. A system (primarily used in residential construction) that incorporates a storage system (batteries) can also provide backup in case the local utility suffers shortages such as those that have occurred in California or due to loss of service as a result of storms, earthquakes, etc.

Wind power is a growing source of electrical generation. In some locations it is beginning to provide a significant amount of electricity. The U.S. is presently the leading producer of electricity from wind ahead of China and Germany. According to the American Wind Energy Association, as of September 2014, there were 46,600 wind turbines operating in the U.S. with a total generating capacity of 62,300 megawatts (MW)—enough to provide power for 15.5 million average American homes. There is currently 13,600 MW under construction across 105 projects (of which 7600 MW is in Texas). The U.S. Department of Energy has set a goal of producing 5 percent of the U.S. electrical requirement from wind power by 2020.

Geothermal Energy

Other green design concepts can be incorporated into a structure. Geothermal design is not particularly difficult to apply and is operationally a simple system. The energy source is free—it comes from the earth (or water) indirectly from the sun. Ground water can be taken from a well or other water source and by use of a heat pump, the energy is then extracted to heat a structure. Underground piping can also be buried in the soil through which a heat transfer fluid is pumped and this absorbs heat from the earth and transfers it to the heat pump. The system does not damage plants or the soil, is endlessly renewable, and can also be used for cooling as well. There are some minor drawbacks, such as the need to move a larger volume of air to achieve the same heating and cooling results, but these problems can be resolved.

Proponents of geothermal systems claim that they can be up to four times more energy efficient as the most efficient fossil fuel systems. Geothermal use can achieve significant savings in heating and cooling and can also produce hot water. Maintenance is reduced as geothermal systems have fewer mechanical components than regular fossil fuel systems. The buried ground loops have a life expectancy of more than 50 years.

Water Harvesting Systems

Significant water savings can be achieved through the use of cisterns. This ancient concept can produce large volumes of gray water for use in watering lawns and garden areas. Two thousand years ago, many Roman homes included cisterns in their courtyards to provide a reliable water source. Modern cisterns are part of what is known as rainwater harvesting systems. These systems consist of three components: a water collection system, a storage cistern, and a distribution method.

There are a number of benefits derived from the use of harvesting systems, including a reduction in the quantity of treated municipal water supplies needed for use in landscaping, laundry, and toilets. In most cases, water used for laundry and sanitary purposes must be eliminated through municipal sewage systems or into a septic system and not used again as gray water. A drawback to the use of harvesting systems includes the potential for significant contamination of collected water due to air and other pollutants. Most collection systems use roof tops as the source of harvested water and channel this by gravity to the cistern. Small pumps then provide for the extraction of collected water and distribution to irrigation pipes, toilets, etc. The cost of dual piping systems needs to be factored into the pricing of these systems. According to most experts, rainwater harvesting systems are often more expensive than municipal water hookups or wells. Balanced against this must be the availability of sufficient water from municipal hookups or wells and the environmental benefit of using all available water sources.

A significant amount of water can be collected using rainwater harvesting systems. For example, an inch of rain over 1,000 square feet of roof generates 600 gallons of water. Put another way, an inch of rain per hour yields about 10 gallons per minute per 1,000 square feet.

Costs/Benefits of Green Buildings

Some environmentalists encourage the design of green buildings without a full understanding of the costs/benefits of this approach and the problems confronted by designers, contractors, and owners. Manufacturers are only now being encouraged to produce green products and major retailers such as Home Depot and Lowe's are just beginning to support the distribution of these products. Much of the available cost/benefit data on the construction of green buildings is incomplete or anecdotal. Additionally, local code officials are often ill equipped to evaluate drawings including green building features and require changes that defeat the purpose of green design.

Current hard data on green building costs/benefits is somewhat limited. In 2003, a report by Gregory H. Kats called *Green Building Costs and Financial Benefits* was published by the Massachusetts Technology Collaborative. At that time, the report noted that "Green buildings provide financial benefits that conventional buildings do not. In particular, energy costs were reported to be on average 30% less in green buildings." I recently installed an array of solar panels on my California residence and have seen a significant drop in electric bills (currently estimated to be 50%). The system was provided at no cost to me by Elon Musk's Solar City on a 20-year lease arrangement. Solar City also provides commercial installations. More recent data provided by the U.S. Green Building Council (published February 23, 2015), notes that "This year it is estimated that 40-48% of new nonresidential construction will be green... [and] 62% of firms building new single-family homes report that they are doing more than 15% of their projects green." As of January 2015, more than 3.6 billion square feet of building space are LEED-certified. Clearly, developers, corporations, home owners, and others are finding green buildings to be highly cost effective.

LEED

The LEED (Leadership in Energy and Environmental Design) Green Building Rating System® was developed by members of the U.S. Green

Building Council. This system is *a voluntary, consensus-based national standard for developing high-performance, sustainable buildings.* LEED standards are currently available or under development for:

- New commercial construction and major renovation projects
- Existing building operations
- Commercial interiors projects
- Core and shell projects
- Homes
- Neighborhood development

In each case, LEED includes the entire design and construction team including architects, engineers, owners, consumers, and all others involved in the process.

According to the U.S. Green Building Council, LEED was created to:

- Define green building by establishing a common standard of measurement
- Promote integrated, whole-building design practices
- Recognize environmental leadership in the building industry
- Stimulate green competition
- Raise consumer awareness of green building benefits
- Transform the building market

LEED is a certification standard intended to distinguish green buildings from the rest of the marketplace. It is a design guideline to encourage sustainability and is also a green building training standard for the construction industry. LEED is organized into five areas:

- Sustainable site planning
- Energy efficiency
- Conserving materials and resources
- Indoor environmental quality
- Water efficiency

For more information on LEED, go to www.usgbc.org.

SELECTED BIBLIOGRAPHY

BOOKS AND MANUALS

American Council of Engineering Companies. *Negotiating for Design Professionals*. Washington, D.C.: American Council of Engineering Companies, 1997.

———. *Project Delivery Systems: The Design Professional's Handbook on Design/Build Project Delivery*. Washington, D.C.: American Council of Engineering Companies, 2001.

———. *Project Collaboration Web Sites for the Design and Construction Industry*. 2nd ed. Washington, D.C.: American Council of Engineering Companies, 2003.

———. *Quality Management Guidelines*. Washington, D.C.: American Council of Engineering Companies, 2003.

———. *ACEC Owner's Project Delivery Manual*. Washington, D.C.: American Council of Engineering Companies, 2004.

American Institute of Architects. *AIA Handbook*. Washington, D.C.: American Institute of Architects, latest edition.

Associated Press. *Style and Libel Manual*. New York: Associated Press, latest edition.

Association for Project Managers. *Prototypical Project Management Manual*. Carmel Valley, CA: Association for Project Managers, 2002.

———. *Project Management Survey*. Carmel Valley, CA: Association for Project Managers, 2003.

Birnberg, Howard. *New Directions in Architectural and Engineering Practice*. New York: McGraw-Hill, 1992.

———. *Project Management for Building Designers and Owners*. 2nd ed. Boca Raton: CRC Press, 1999.

Burford, Laura. *Project Management for Flat Organizations*. Plantation (FL): J. Ross Publishing, 2013.

Burstein, David, and Frank A. Stasiowski. *Total Quality Project Management for the Design Firm*. New York: Wiley, 1993.

Cappels, Thomas M. *Financially Focused Project Management*. Plantation (FL): J. Ross Publishing, 2003.

Cleland, David I., and Lewis R. Ireland. *Project Management*. 4th ed. New York: McGraw-Hill, 2002.

Clough, Richard H., and Glenn A. Sears. *Construction Project Management*. 4th ed. New York: Wiley, 2000.

Cooper, Dale F. *Project Risk Management Guidelines*. New York: Wiley, 2004.

DeCarlo, Doug. *Extreme Project Management*. New York: Wiley, 2004.

Englund, Randall L. *Creating the Project Office*. New York: Wiley, 2003.

Forsberg, Kevin. *Visualizing Project Management*. 3rd ed. New York: Wiley, 2005.

Garrett, Dave, ed. *Project Pain Relievers: A Just-In-Time Handbook for Anyone Managing Projects*. Plantation (FL): J. Ross Publishing, 2012.

Gray, Colin, and Will Hughes. *Building Design Management*. Oxford: Butterworth-Heinemann, 2001.

Ingardia, Michael P., and John F. Hill. *Contracting for CADD Work: A Guide for Design Professionals*. Washington, D.C.: American Council of Engineering Companies, 1997.

Kerzner, Harold. *In Search of Excellence in Project Management*. New York: VNR, 1998.

———. *Advanced Project Mangement*. 2nd ed. New York: Wiley, 2003.

———. *Project Management: A Systems Approach to Planning, Scheduling and Controlling*. 9th ed. New York: Wiley, 2005.

———. *Project Management Case Studies*. 2nd ed. New York: Wiley, 2006.

Klastorin, Ted. *Project Management: Tools and Trade-offs*. New York: Wiley, 2003.

Levin, Ginger, and Parviz Rad. *Achieving Project Management Success Using Virtual Teams*. Plantation (FL): J. Ross Publishing, 2003.

Levy, Sidney M. *Project Management in Construction*. 4th ed. New York: McGraw-Hill, 2002.

Mantel, Jr., Samuel J. *Core Concepts*. 2nd ed. New York: Wiley, 2004.

Meredith, Jack R., and Samuel J. Mantel, Jr. *Project Management: A Managerial Approach*. 6th ed. New York: Wiley, 2005.

Moustafaev, Jamal. *Delivering Exceptional Project Results*. Plantation (FL): J. Ross Publishing, 2011.

Newbold, Robert C. *Project Management in the Fast Lane*. Boca Raton: CRC Press, 1988.

Nicholas, John M. *Project Management for Business and Engineering*. 2nd ed. Burlington: Elsevier, 2004.

O'Brien, James J. *CPM in Construction Management*. 4th ed. New York: McGraw-Hill, 1993.

Phillips, Jack J. *The Project Management Scorecard*. Burlington: Elsevier, 2002.

Project Management Institute. *Project Management Book of Knowledge*. latest edition, Newtown (PA).

Rakos, John. *The Practical Guide to Project Management Documentation*. New York: Wiley, 2004.

Rojas, Eddy, ed. *Construction Project Management*. Plantation (FL): J. Ross Publishing, 2009.

Rose, Kenneth H. *Project Quality Management*. 2nd ed. Plantation (FL): J. Ross Publishing, 2014.

Rosenau, Milton D., and Gregory D. Githens. *Successful Project Management*. 4th ed. New York: Wiley, 2005.

Stephenson, Jr., Ralph. *Project Partnering for the Design and Construction Industry*. New York: Wiley, 1996.

Stone, David A. *Hard-Core Project Management*. Raleigh: FMI, 1999.

Teicholz, Eric. *Facility Design and Management Handbook*. New York: McGraw-Hill, 2001.

Tunstall, Gavin. *Managing the Building Design Process*. Oxford: Butterworth-Heinemann, 2000.

United States Department of Veterans Affairs. *A/E Quality Alert*. Washington, DC, 1997.

Verzuh, Eric. *The Fast Forward MBA in Project Management*. 2nd ed. New York: Wiley, 2005.

Warne, Thomas. *Partnering for Success*. Washington, DC: American Society of Civil Engineers, 1994.

Wysocki, Robert K. *Effective Project Management*. 2nd ed. New York: Wiley, 2000.

OTHER SELECTED RESOURCES

American Council of Engineering Companies (ACEC), 1015 15th Street NW, Washington, D.C., 20005; 202-347-7474.

American Society of Civil Engineers (ASCE), 1801 Alexander Bell Drive, Reston, VA 20191; 703-295-6252/800-548-2723.

American Society for Personnel Adminstration/Society for Human Resource Management (SHRM), 1800 Duke Street, Alexandria, VA 22314; 703-548-3440/800-283-7476.

American Society for Training and Development (ASTD), 1640 King Street, P.O. Box 1443, Alexandria, VA 22313; 703-683-8100.

ASFE: Professional Firms Practicing in the Geosciences (ASFE), 8811 Colesville Road, #G106, Silver Spring, MD 20910; 301-565-2733.

Association for Project Managers (APM), 288 El Caminito Road, Carmel Valley, CA 93924; 312-664-2300.

Construction Specifications Institute (CSI), 99 Canal Center Plaza, #300, Alexandria, VA 22314; 703-684-0300/800-689-2900.

Harvard Design School, Office of Executive Education, 1033 Massachusetts Avenue, 5th Floor, Cambridge, MA 02138; 617-496-0436.

Project Management Institute (PMI), 14 Campus Blvd., Newtown Square, PA 19073; 610-356-4600 (US), customercare@pmi.org.

Institute of Management and Administration, 29 West 35th Street, New York, NY 10001; 212-244-0360 (website http://www.ioma.com).

University of Wisconsin-Extension Engineering, 432 North Lake Street, Madison, WI 53706; 608-262-2061.

ARTICLES AND REPORTS

"A Mighty Wind is Blowing Across the Land." *Architectural Record*, May 2005, 273–278. Barbara Knecht, author.

"After Theory." *Architectural Record*, June 2005, 71. Michael Speaks, author.

"Award of Excellence-Charles Thornton." *ENR*, April 23, 2001, 30. Nadine Post, author.

"Building Costs and Financial Benefits." Massachusetts Technology Collaborative, 2003. Boston, MA. Gregory Kats, author.

"Case History No. 70: In the Real World of Geoprofessional Practice." Professional Firms Practicing in the Geosciences (AFSE). Silver Spring, MD, 1996.

"Listen Up or Lose the Client." *Architecture*, February 1996, 153. Barry LePatner, author.

"Looking at Where Architects Are." *Architectural Record*, February 2004, 70. Charles D. Linn, FAIA author.

"Make Peace with Grammar and Punctuation." *CE News*, November 2000, 54. Cathy Murphy, author.

"Negotiator at Large-Bill Richardson." *New York Times Magazine*, January 26, 2003, 13. David Wallis author.

"Preparing and Editing Written Materials." *Project Manager* (APM Journal), Winter 2003, 27. Howard Birnberg, author.

"Research Suggests." Chicago Tribune, March 2, 1998, (business) section 4, 2. Jon Van and Jon Bigness, authors.

"Risk Management for Owner and Design Firm Project Managers." *Project Manager* (APM Journal), Fall 2001, 13. Howard Birnberg, author.

"Seismic Performance Analysis and BIM." *CENews.com*, January 2015, 30. Leo Salce and Joshua Gionfriddo, authors.

"Speak Out," *Architectural Record*, November 1998, 24. Ava Abramowitz, author.

"Web Collaboration Tools-Will They Fit On the A/E/C/ Industry Tool Belt?" *Civil Engineering News*, April 2001. Robert S. Schneider, author.

"Why Architects Don't Charge Enough." *Architectural Record*, October 1999, 10. Elizabeth Harrison Kubany and Charles D. Linn, AIA authors.

INDEX

Accelerated decision-making process, 243–245
Account managers, 12
Active encouragement, 63. *See also* Mentoring programs
Activity ground rules, 221
A/E/C Systems, 235–236, 240–242, 245–246
Allowances, 207, 209
Alternative dispute resolution (ADR) devices, 190
American Council of Engineering Companies (ACEC), 55, 130, 197
American Institute of Architects (AIA), 55, 95, 117, 130, 151, 197, 233
American Society for Testing and Materials (ASTM), 206
American Society of Heating, Refrigerating and Air-Conditioning Engineers (ASHRAE), 246
Association for Project Managers (APM), 31
Association of Engineering Firms Practicing in Geosciences (ASFE), 197

At-risk CMs, 19. *See also* Construction manager (CM)

Bar charts, 218
Better Tires and Transmissions, 88
Bidding process, 128
Billing and collection, 156–158
BIM Software, 142, 247–249
Brooks Act, 107
Budgeting project design costs, 99–102
  direct personnel expense, 105–106
  plan, 102–105
Building Design Management, 121
Building Information Modeling (BIM), 28–29, 236–237

Client selection of design consultants, 12–13
Computer-aided design and drafting (CADD), 233
  CADD Systems, 237–240
  capability, 234
  drawings, 191
  training, 52
Computer applications
  case study, 249–250
  in 1950s, 231–232

in 1960s, 232–233
in 1970s, 233–234
in 1980s, 234–235
in 1990s, 235–236
in 2000s/2010s, 236–237
Computerized Financial Management System (CFMS), 233
Construction administration, 129
Construction manager (CM), 19, 128–129
Construction Specifications Institute, 178
Construction Specifier magazine, 197
Consultant work authorization form, 168
Contract management/project administration, 151
   administrative activities, 174–178
   billing and collection, 156–158
   case study, 162–169, 182–184
   contract types, 154–156
   design budget, 172–174
   financial issues, 152–153
   notebooks and manuals, 179–182
   partnering, 184–185
   payment, 158–161
   project design costs and schedules, 169–171
Contractors, 70
Contract staffing
   contract staff, 69–70
   employers, 71
Contract types. *See also* Contract management/project administration
   design services contracts, 154–156

Cost control
   communicating with clients, 202–203
   estimating and controlling, 203–205
   information systems, 203
   internal project, 201–202
Costs and schedules, 169–171
Critical path method, 218–224
Cross training, 57–58

Delivery methods
   design/build method, 19–20
   fast-track method, 18–19
   traditional straight-line method, 16–18
Delivery system, 7
   design-build, 20
   types, 8–12
Deltek Software, 233
Departmental organization, 9–10
Design and construction (D&C) operations, 47
Design/bid/build system, 242
Design/build method, 19–20. *See also* Delivery methods
Designer/client relationships
   client retention, 135–136
   owner/client expectations, 136–144
Design firm operations
   budgeting, 99–106
   case study, 106–113
   profit planning, 91–95
   salary information, 113–115
   scope determination, 95–97
   selecting external consultants, 97–99
Design firm project management, 2–3

case study, 20–22
Design process managing
　case study, 117–119
　changing ideas, 119–120
　engineering design, 124–125
　evaluating and selecting designers, 120–123
　parts, 123–124
Design report card form, 252
Design services billing checklist, 159
Design services contracts
　cost-plus-fixed-fee, 155
　lump-sum, 154–155
　multiplier-times-salary, 155–156
　percentage-of-construction-cost, 156
　time-and-materials, 154 (*See also* Contract types)
　value-of-service, 156
Direct personnel expense (DPE), 105–106
Dummy variable, 221

Engineering design, 124–125
Engineering News-Record (ENR), 246–250
Engineering Times, 69
Engineers Joint Contract Documents Committee (EJCDC), 151
External constraints, 122–123
External consultants
　owner/facilities managers, 99
　selection process, 97–99

Facilities project managers, 3
Fast-track method, 18–19. *See also* Delivery methods

Federal Acquisition Regulations (FAR), 107
Flexible project delivery systems, 242
Formal mentoring programs, 62–63
Full wall scheduling, 218
Function analysis, 211–214
Function analysis system technique (FAST), 212

Gantt charts, 218–221
General Services Administration (GSA), 109
Getting There by Design: An Architect's Guide to Design and Project Management, 29
Goal setting
　bottom-up decision making, 25
　incremental or continual decision making, 25–26
　pragmatic compromise, 25
Great Recession, 1–2
Green Building Costs and Financial Benefits, 259
Green Building Rating System®, 259–260
Green building revolution, 256–257. *See also* Project environmental considerations

Hire/fire short-term staff, 68

Impact on construction costs, 12
Independent contractors, 70
Informal mentoring programs, 63
Information for completed project file (electronic and/or paper), 101

Information modeling, 245–249
　data retention, 251–253
　legal concerns, 253
　management and marketing use, 254
Information Strategy magazine, 28
Instant minutes, 79–80
Integrated Project Delivery (IPD), 247–248
Internal constraints, 122
International Facilities Management Association (IFMA), 55
Intern Development Program (IDP), 60–61

Kismet, 63. *See also* Mentoring programs

Leadership in Energy and Environmental Design (LEED), 259–260. *See also* Project environmental considerations
Long-range planning process, 24–25

Management concepts
　decision making, 26–29
Manuals
　content, 180–182
　development, 180
Marketing and design firm project managers, 40–42
Matrix management, 10–12
Mechanical-electrical-plumbing engineer (MEPE) projects, 247–248
Meeting management, 79–80. *See also* Soft skills for PM

Mentoring programs, 59
　types, 62–63
Mentoring project managers. *See also* Project managers (PM)
　mentoring programs, 61
Michigan Department of Transportation (DOT), 70
Multidiscipline firms, 10

National Society of Professional Engineers (NSPE), 55, 197, 233
National Society of Professional Engineers (NSPE) publication, 69
Negotiating skills for PM, 88–90
Network, 224–225
　limitations, 225
　phases, 220
　sample, 222
Not-at-risk CMs, 19. *See also* Construction manager (CM)
Notebooks, 179–180

On-the-job training (OJT), 59
Overburdened senior managers, 5
Owner program management, 144
　delivery options, 146–147
　issues in selecting delivery options, 145
　owner capabilities, 145–146
　owner requirements, 145
　owner responsibilities, 146
　program management, 147–149

Partnering, 184–185. *See also* Contract management/project administration
Partnering for Success, 184

Peer review, 197–200
Performance-based codes and standards, 209
Personnel planning, 65–66
Personnel planning and management, 65–67
  contract staffing, 69–71
  leveling workload, 68–69
Phases and personnel responsibilities, 127–129. *See also* Staff and responsibilities
  designer/client relationships, 135–144
  owner program management, 144–149
  staff, 130–135
Planning
  concepts, 23–24
  issues, 227–228
Principal or partner in charge (PIC), 130
Profit planning for design firms, 91
  labor, 92
  non-labor costs, 92–94
  ratios/multipliers, 94–95
Program evaluation and review technique (PERT), 219
Program management, 147
  applicability, 149
  contractual relationships, 148
  scope of services, 148–149
Program of Requirements (POR), 107–108
Project administrative activities
  checklists, 177–178
  check prints, 178
  confidentiality, 177
  dealing with consultants, 177
  e-mail/text messages, 178
  filing project paper data, 174–175
  incoming correspondence, 175
  interoffice/intraoffice memos, 175–176
  manufacturer's assistance, 176–177
  meeting notes, 176
  outgoing correspondence, 175
  telephone calls, 176
  work authorizations, 176
Project environmental considerations, 255–256
  costs/benefits of green buildings, 259
  electrical generation, 257
  geothermal energy, 257–258
  green building revolution, 256–257
  LEED, 259–260
  water harvesting systems, 258
Project management
  in small design firms, 3–4
  system, 5–7
Project managers (PMs), 3
  areas requiring attention, 36–38
  case study, 34–36, 46–48
  characteristics, 32–33
  communication skills, 33
  finding, 42–44
  keeping, 44
  and marketers, 254
  need, 44–46
  responsibilities, 38–40
  rewards, 48–49
  roles, 29–30
  training, 42–44, 50–56
Project team defining, 4
Public speaking techniques for PM, 82–87

Pyramid approach, 8–9

Qualification-based selection (QBS) process, 120–121, 138–139
Quality and risk management, 187
  documentation, 191–192
  effective communication, 189–191
  total approach, 188–189
Quality assurance program, 194
  steps in developing, 195–197

Red flag words, 189–191
Reimbursable markups, 114–115

Scheduling
  benefits and limitations, 224–226
  issues, 226, 228–229
  methods, 217–224
Scope determination by design firms, 95–96
  dividing contracts, 97
Scope management, 162–164
  communicating design change orders, 166–169
  design contract change orders, 165
Scope of services planning form, 96
Seismic Performance Analysis and BIM, 248
Self-employment, 70
Single-discipline departmental system, 9–10
Site administration and information control, 242–243

Slack time, 222
Soft skills for PMs
  meeting management, 79–80
  telephone time management, 80–81
  time management, 77–78
Specialists and generalists, 133–135. *See also* Staff and responsibilities
Specialized consultants, 132–133. *See also* Staff and responsibilities
Specifications and construction drawings, 205–206
  allowances, 207–209
  performance-based codes and standards, 209
Specification writers, 132
Staff and responsibilities
  position descriptions, 130–132
  specialists and generalists, 133–135
  specialized consultants, 132–133
Staff development officer, 56–57
Staff management
  delegating responsibility, 74–75
  responsibility and authority, 74
  techniques for proper delegation, 72–74
  training, 52–53
Staff productivity, 68
Strategic project management plan, 13
  elements, 14–15
  items, 15–16
STRESS Language, 232
Studios, 12

Telephone time management, 80–81. *See also* Soft skills for PM
Telephone/voice mail lesson, 82. *See also* Soft skills for PM
Time management, 77–78. *See also* Soft skills for PM
Total quality management, 193–194
Traditional straight-line method, 16–18. *See also* Delivery methods
Training practices and methods, 58–59
Training program, 52
 effective learning, 57
 evaluation, 60
 managing, 60
 need, 53–56
 sources/providers, 60–61

Uniform Building Code (UBC), 143–144

United States Internal Revenue Service (IRS), 71

Value engineering (VE), 209
 customers, 211
 function analysis, 211–214
 implementation results, 210–211
 job plan, 214–215
Value pricing and marketing, 113–114

Weak/ineffectual project management systems, 5–7
Web collaboration potential, 245
Web collaboration tools, 240–245
Work authorization form, 166–167
Work overtime, 68
Written materials, preparing and editing, 82–84